INTRODUCTION TO ELECTRONIC SPEECH SYNTHESIS

by
Neil Sclater

Howard W. Sams & Co., Inc.
4300 WEST 62ND ST. INDIANAPOLIS, INDIANA 46268 USA

FIRST EDITION
FIRST PRINTING—1983

International Standard Book Number: 0-672-21896-8
Library of Congress Catalog Card Number: 82-60872

Edited by: Frank N. Speights
Illustrated by: T. R. Emrick

Printed in the United States of America.

Preface

A knowledge of the human physiology of speech generation may not be essential for the hobbyist and those circuit designers interested only in experimenting with speech synthesis devices and the development kits that are based on existing hardware, but it is essential to the product or system designer who is looking beyond the existing demonstration kits into synthesis-by-rule, text-to-speech, and even speech recognition concepts that are still under active development. Also, complex mathematics is needed for a complete explanation of linear predictive coding (LPC), which is one of the important techniques now being applied to speech synthesis. However, this math has been omitted from this book.

This book is written as a primer in speech synthesis for the person without formal training in computer hardware, computer programming, or the physiology of speech. Speech synthesis is a marriage between digital electronics (primarily microcomputer) and linguistics (the science of language). Thus, this book is intended to answer the question: "What is being done in the field of electronics, with readily available integrated circuits and board-level products, that could be used to provide a synthesized human-voice output from present or future products?" The reasons why one would wish to add this feature to a product run the gamut from entertainment and education to providing hazard warnings and aids for the handicapped.

The digital electronic filter that is built into a number of different synthesizer integrated circuits is a microminiaturized analogy of the human vocal tract and is discussed in detail. The adaptation of the complex multistage filter into a low-cost silicon chip is one of the "breakthroughs" that has turned electronic speech synthesis into a viable technology suitable for many cost-effective commercial applications.

LPC is not the only approach to electronic speech synthesis. Other methods also depend on the recording and the conditioning of the actual sounds of the human voice but they arrange the frequency components to permit a more efficient speech compression, thereby reducing the amount of memory needed to store data for speech reconstruction.

In all forms of synthesis, the restoration of recognizable speech is a little like adding hot water to freeze-dried coffee crystals. The digital data stored in memory corresponds to the coffee crystals and the synthesizer circuit provides the "right amount" and "right temperature" of hot water to make what approaches a cup of fresh coffee.

The current wave of interest in speech synthesis was initiated by product introductions from two major American semiconductor manufacturers—Texas Instruments Incorporated and National Semiconductor Corp.—whose approaches are distinctly different. Texas Instruments has pursued a frequency domain synthesis approach while National Semiconductor has pursued a time domain synthesis approach.

Both companies offer synthesizer chips, memories, and even demonstration kits that are relatively low in cost. These kits, available with on-board semiconductor memory synthesizer, battery, speaker, and a few switches, will adequately demonstrate (without the need for a microcomputer) synthetic speech that is arrived at by differing methods. Although some of these kits are described and illustrated in the following chapters, this book does not include the "how-to" aspects of setting up and operating these boards. The manufacturers' applications literature available with those products is a more than adequate source for that information. However, since there is no standardization or uniformity in these products, it should be noted that the "how-to" instructions will probably differ for each product.

Other companies that have introduced integrated-circuit-level devices and some demonstration hardware include American Microsystems, Inc. and General Instrument Corp., but their devices are patterned after the Texas Instruments' concepts. Because of the dynamic nature of this marketplace, no attempt has been made to catalog all the U.S. and foreign companies, particularly the Japanese, that have announced their participation. There are wide discrepancies in the amount and quantity of background literature and applications notes available from these companies, suggesting significant differences in dedication to the subject and in the level of participation, as well as the strategy in the marketplace. Most semiconductor firms are interested in electronic speech synthesis on two levels: (1) the immediate sale of their proprietary products and services, and (2) the encouragement of the tie-in sale of their microcomputers and other semiconductor products.

There is yet another group of companies participating in the electronic speech synthesis marketplace—the system integrators with an end-use application orientation. These companies offer demonstration modules and products, incorporating their own approaches to speech synthesis, directly to industrial and commercial original equipment manufacturers (OEMs). These products are generally tailored to meet a specific need. These firms are not as well known as the semiconductor houses but they appeal to the OEMs who see a need for electronic speech synthesis in their products but lack the resources or personnel to develop a system from the chip-level devices. The applications companies serve the dual role of consultants and system integrators for their customers. Several of these companies have designed and developed their own silicon integrated-circuit synthesizers, which were then made for them by custom specialists, but they are not limited to these devices.

The whole field of electronic speech synthesis at the commercial chip-and-board level is so new that there is no second sourcing of products by any of the semiconductor manufacturers and there is no compatibility between them. The subject is also complicated by a lack of uniformity in the use of terminology in the literature and applications notes. Most of the available literature is organized in the format of semiconductor specifications sheets, but the symbols and descriptive terms vary. This is frustrating to the average reader who is concerned with drawing inferences about performance from the available literature.

This book is intended as a technical but nonmathematical introduction to the current state-of-the-art in speech synthesis for commercial, consumer, and industrial applications. I have attempted to reconcile the variations in terminology, which is a mixture of terms that have their origins in both linguistics and in microprocessor hardware and software. As an additional aid, a glossary of electronic speech synthesis terminology is given in Appendix A.

NEIL SCLATER

Contents

APPENDIX E

APPENDIX F

CHAPTER 1

WHAT IS SPEECH SYNTHESIS?

Electronic speech synthesis (ESS) is in its formative years and, like many an adolescent, is experiencing growing pains. No single approach has emerged that is universally acceptable to meet the many applications seen for ESS. As a result, there are many participants, semiconductor manufacturers as well as consultant/system-integration firms, that are "going-it alone" in seeking to convince circuit and system designers that their approach to synthesizer and memory chips, board-level products, and supporting services has the most to offer for the money. At this time, there is no second sourcing or fabrication of compatible speech synthesis devices by competitive semiconductor manufacturers, and this is a widely accepted precursor to volume sales of any electronic component.

A number of different methods have been developed for the reproduction of the human voice from stored digital data, which is the universally accepted common denominator in ESS today. These methods will be described in this book in enough detail to permit the reader who is not an engineer to understand the differing concepts. There are many options open to the OEM who wishes to add voice output to his product or system. All are closely tied to microprocessor technology and they draw on the techniques and components that have proven successful with those devices.

It is expected that ESS will contribute significantly to the growth of the electronics industry over the next decade and will spearhead the advance of electronics into many new and different applications. It is also safe to assume, based on the history of other semiconductor devices, that as the price of ESS components falls due to learning-curve increases in productivity, many applications now closed due to cost considerations will open.

ESS AVAILABILITY

Speech synthesis devices that will permit the playback of a limited "canned" vocabulary of stock words and phrases can be purchased today at a very low cost. The speech quality of these devices is surprisingly good but the cost per word or phrase is prohibitively high for many potential applications. By contrast, with a larger investment, it is possible to buy the devices and board-level systems that will permit the concatenation of various sounds or word elements necessary to form realistic and understandable words and phrases. The costs of the two approaches may be orders of magnitude apart, but the cost to produce a word or phrase will be considerably lower.

At this stage in the development of ESS, one can say that if the objective is to electronically synthesize a limited vocabulary of a dozen words or so of standard speech, implementation with the products of a number of vendors is simple and inexpensive. However, if the object is a number of complete sentences while still using a standard vocabulary, price will rise and there is a good chance that the speech quality will be lower. Going a step further, if the user would like a true custom speech or vocabulary, the cost will be considerably higher and there will probably be service charges and a delay in hardware delivery.

If an OEM wants to prepare his own words and phrases in his own facilities, the equipment is available. But to do so, he must make a commitment to the semiconductor products and technology of a specific supplier. And if an OEM should be content with the reproduction of just a few words or phrases (with no concern for future expansion), it would be safe to say that the economics of using ESS may not be much more favorable than using some form of low-cost audio-tape playback machine—as is being done in one present-day Japanese automobile. However, in addition to the larger bulk of a closed-loop tape-playback machine, that system is also handicapped by a higher power consumption, and an inherently lower reliability—two more drawbacks redressed by ESS.

FUTURE OF ESS

Work is already underway for new and improved generations of speech synthesizer chips. Moreover, work is also progressing in improved methods for speech compression to reduce further the cost of data storage, the pivotal expense in any of the approaches. Is speech quality best improved at the memory or data-storage level, or is it best improved by adding features to the synthesizer chip? This question lies at the heart of some of the differing concepts being applied to speech quality.

The prime candidates for ESS today are those products or systems in which an electronic voice response would add true value. Whether voice announcements would be initiated manually or work automatically from appropriate sensors is one of the design decisions. A voice response feature can be used to issue instructions, offer guidance, or elicit warnings pertaining to personal safety or protection of property. The following list of applications suggests some areas now under investigation or those for which prototypes have actually been developed:

- Elevators that announce floor numbers and the merchandise located on each floor.
- Automobiles that issue safety warnings or advise on malfunctions or the need for maintenance.
- Banking terminals that offer step-by-step instructions in deposit and withdrawal procedures. These are especially useful to those persons with impaired vision or a minimal reading ability.
- Factory assembly or test stations that provide voice guidance or instruction for equipment operation with the intent of reducing errors and increasing productivity.
- Automated teaching machines that permit students to interact with a computer during drills on language, spelling, mathematics, and other subjects.
- Automated public-address systems in public buildings or public transportation that are programmed to issue appropriate orders for evacuation or self-protection in the event of fire, power outage, nuclear plant malfunction, or other emergency situations.
- Prosthetic devices for persons with a temporary or permanent speech disability. These devices might also be of value in summoning aid for the deaf, the aged, or the mentally retarded.

Voice-response systems may be simple stand-alone modules requiring only a speaker and a battery, or they may be more elaborate circuits that are tied into telecommunications or computer systems. Needless to say, there is a real danger that low-cost systems could be built into toys and novelties having no educational or other redeeming feature, and they could become nuisance contributors to "noise" pollution. A clock that "speaks" the time could be of immense value to the blind or to the handicapped, but it could be a source of annoyance to people with normal faculties.

The quality of the voice output has been the subject of heated debate with some arguing that a metallic-sounding monotone "robot" voice may, in fact, have more attention-grabbing value in a public-address system—provided it was understandable—than a more natural-sounding voice that might be confused with a radio announcement. On the other hand, continual instructions or messages in that type of voice could also prove to be an irritant except under emergency or unusual conditions.

It has been estimated that more than a billion dollars will be spent on ESS hardware by the end of the decade. However, most observers see the future of ESS intertwined with that of an allied but independent technology—electronic speech recognition, or ESR. While speech recognition will not be covered in this book, it is worth pointing out the fact that speech synthesis is much further ahead in terms of the development and availability of hardware.

SPEECH RECOGNITION

Recognition of speech is far more complicated than speech synthesis because of the difficulty in building a circuit that will respond to the wide variations of acceptable human speech in any one language. While a number of different speech recognition systems have been developed, none are truly universal. They must be "trained" to accept the vocal characteristics of a particular speaker and they are limited to the "comprehension" of a relatively small number of words that are spoken in a prescribed manner. Obviously, the ability to recognize and respond to a specific speaker has some commercial value in products, such as security systems, that are intended to be selective. However, this is a drawback for wide applications involving a number of different speakers.

There will be many improvements in ESR in the future as better and lower-cost methods of analyzing and comparing the root sounds, or *formants*, of speech are introduced. Many of these improvements might themselves be the "fallout" from research and development in speech synthesis. To what extent standardization of ESS technology might impact ESR is still an open question.

A marriage between ESS and ESR systems would open the door for truly interactive computer conversations. It is entirely possible that translating human messages into a "uniform" synthesized voice would make it easier for the recognition system to function. It is also possible to visualize, on a basis of existing technology, "conversations" between robots that could be understood by human listeners. While fascinating to contemplate, these subjects quickly take on the aura of science fiction and that is not the purpose of this book. Emphasis is placed on existing hardware that is available on order and, in some cases, available off-the-shelf in retail stores and from electronics component distributors. The pace of the speech synthesis field is so rapid that it was necessary to exclude the mention of some products now known to be under development. Announcement of future products would violate the "here and now" emphasis in this book.

CONVERSION METHODS

Techniques that permit the conversion of speech to a digital format for storage or transmission have been known for years. Analog-to-digital conversion and pulse code modulation (PCM) are two methods that use uncompressed digital data. They are the simplest to use and understand. These methods permit the conversion of speech signals into digital data using an analog-to-digital converter. Once in digital format, the signals can be stored in any of a number of different kinds of memory and can be played back through a digital-to-analog converter. A low-pass filter is used to smooth out the signal and reduce noise prior to amplification so that it can drive a speaker. One shortcoming of this approach is the large amount of memory required to store the signal. Magnetic tape or disks are frequently used with PCM today.

Signal Sampling

A voice signal is converted to a digital signal by a method known as sampling. The analog signal is sampled periodically and often enough to permit the original analog signal to be reconstructed. Typically, the voice signal is sampled at a minimum rate of twice the highest frequency that is present in the signal. Since the effective upper frequency of the human voice is about 4000 Hz, the minimum sampling rate would be twice that, or 8000 samples per second.

The variations in amplitude of the human voice can be crudely represented by a 4-bit number. However, if 8 bits or even 12 bits are used, the quality or reproduction can be significantly improved. Unfortunately, the number of bits chosen to represent each speech sample influences the data rate and the amount of memory required to store the digitized speech. The number of bits chosen is multiplied by the sampling rate to yield the data rate in bits per second. For 8 bits, this would be 64,000 bits per second and for 12 bits it would be 96,000 bits per second. In either case, this calls for more memory at a higher cost than would be acceptable for most commercial applications. Some of the largest random-access memories that are commercially available today contain 65,536 bits—enough for one second of speech when using only 8 bits.

Signal Storing

There are many advantages in converting the human voice to a digital format for transmission in various communications systems. However, not all these applications have a requirement for storing the digitized signal. Nevertheless, if it is desirable to store the signal, even temporarily, a technique known as speech compression may be useful in reducing the amount of storage required or, conversely, increase the length of speech that can be stored on a particular kind of memory.

SPEECH SYNTHESIS

Speech synthesis, which includes speech analysis, calls for determining the operating rules and techniques for reproducing speech. It was originally motivated by a desire to improve the efficiency of digital communications. In the past, scientists have studied the human voice and tried to reproduce it artificially using different mechanical and electrical models. But serious work on speech synthesis was not attempted until the digital computer was developed. This led to true electronic speech synthesis (ESS).

Some forms of ESS take advantage of the predictability and redundancy of speech and they provide an output code that is interpreted by the human ear as speech. Other methods reproduce speech by reconstructing the original waveform from a knowledge of changes in pitch, energy, and other characteristics.

There are three different approaches to electronic speech synthesis today: analysis synthesis, constructive synthesis, and hybrid synthesis. Each of these approaches will be discussed in further detail later in this book and some examples of practical hardware will be described. Unfortunately, at this stage in the development of ESS, there are strong commercial biases aimed at promoting one approach over the other. The present hardware manufacturers have made extensive investments in the development of certain kinds of semiconductor integrated circuits embodying one type of approach and they are anxious to obtain a reasonable return on their investments.

Each manufacturer of speech synthesis devices is likely to include certain proprietary variations on a recognized general approach. For obvious reasons, their advertising and promotional literature tends to stress the supposed advantages of their approach to the exclusion of others. While there are genuine differences in analysis synthesis approaches today, they are more likely to be significant in the cost of hardware and speech preparation than in the final voice output. Discrimination of quality between the simple low-cost schemes that are readily available today can be quite difficult for the average listener when the synthesized speech is limited to simple words, phrases, or even sentences.

However, the present situation of one company/one technology is rapidly changing and one can expect that the larger suppliers, at least, will soon be offering a choice of approaches where the selection of an approach will be directly related to the application of the synthetic speech and the amount of money that the customer is willing to pay for such an installation.

ANALYSIS SYNTHESIS

Analysis synthesis is a term used for methods that derive synthetic speech from the recording and compression of the actual human voice. There are two general approaches used today: (1) time domain synthesis which yields a synthetic waveform representation of the original speech, and (2) frequency domain synthesis which yields a parametric representation of the frequency spectrum of the original speech.

Time Domain Synthesis

In time domain synthesis, a synthetic speech waveform is constructed which sounds very much like the original waveform but which looks very different. About half of the synthetic waveform is silence. The waveform is made up of many symmetric segments which range over a very restricted set of amplitude values. Because of these features, the synthetic waveform can be described and it can be stored using about

1% of the bits that are required to reproduce the original speech. The synthetic waveform is constructed by adjusting the phases of all the frequencies in the original signal while keeping the amplitudes (the power spectrum) as close as possible to the original.

Frequency Domain Synthesis

There are two general approaches being used in frequency domain synthesis today. Both approaches make use of parametric encoding of the (frequency) power spectrum of the speech waveform. They are *formant synthesis* and *linear predictive coding (LPC).* In the parametric encoding schemes, speech characteristics other than the original waveform are used in the analysis and synthesis of speech. These characteristics are used to control a mathematical synthesis model to create an output speech signal that is similar to the original. The speech output waveform of a frequency domain synthesis signal does not necessarily resemble the original speech signal introduced into the input analyzer. However, the objective here is to duplicate the spectral shape of the speech signal as it is formed by the human vocal tract. The speech output waveform need not be a reasonable match for the input waveform.

Formant synthesis is a frequency domain synthesis technique that reproduces speech by recreating the spectral shape of the waveform. It uses the *formant* center frequencies, their bandwidths, and the pitch periods as inputs. A formant is a frequency region where the energy of the vowel sound is concentrated. Formants show up as frequency peaks in the voice spectrum.

The second important frequency domain synthesis encoding technique, *linear predictive coding* or *LPC*, is based on a mathematical model of the human vocal tract. Pitch and energy information and speech variables used to model the vocal tract are obtained from speech recordings. The speech data are analyzed and encoded to produce input data suitable for the digital model. The LPC method of speech analysis accurately preserves the character of the speaker's voice, including intonation, accent, dialect, and pitch in any language.

Memory Requirements

Fig. 1-1 is a diagram that illustrates the general categories of analysis synthesis as compared with pulse code modulation (PCM). The diagram indicates some of the tradeoffs that apply. PCM is the easiest and lowest-cost method for digitizing human speech and is widely used in

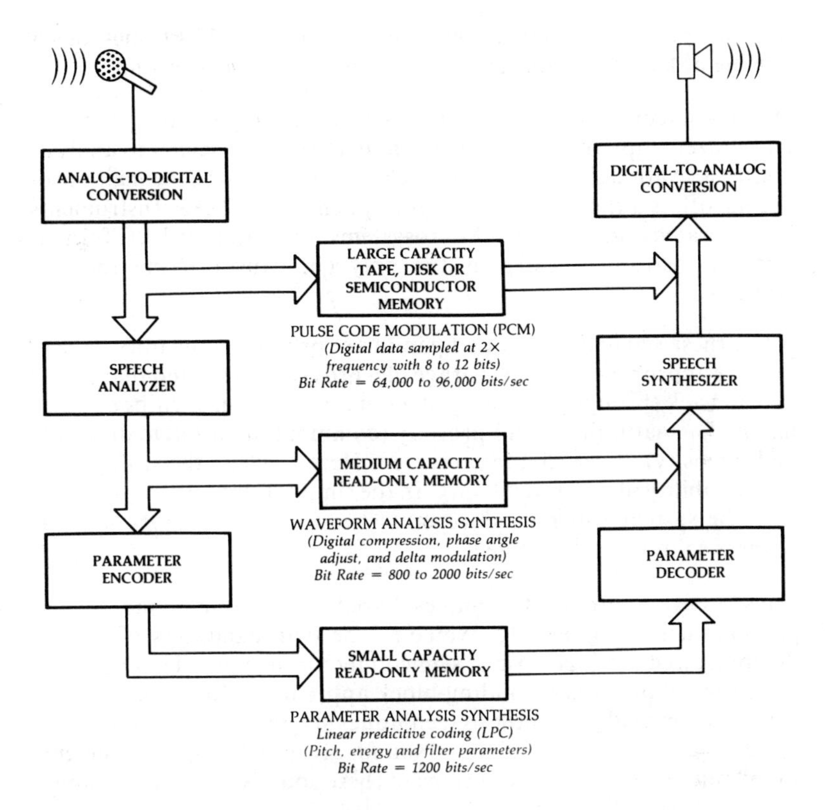

Fig. 1-1. Comparing pulse code modulation to waveform and parameter analysis synthesis encoding techniques.

telecommunications but its use in electronic speech synthesis would call for the largest amount of memory.

The speech compression applied in time domain synthesis requires more costly processing equipment than what is needed for PCM but it offers a reduction in data rate and a savings in the amount of memory needed to store a given amount of speech. The data rate of the most widely used time domain synthesis encoding technique is approximately 1800 bits per second.

To carry out the frequency domain techniques, much more sophisticated synthesis encoding equipment is needed. For example, the equipment employed for formant synthesis is typically able to encode at

1200 bits per second while the equipment used for the LPC methods is capable of handling data rates of 1200 to 2400 bits per second.

Major semiconductor manufacturers now offering electronic speech synthesizer chips are not in agreement over the approaches to analysis synthesis. National Semiconductor Corp., alone among the major producers, offers a time domain analysis system, while Texas Instruments Incorporated and American Microsystems, Inc. offer the LPC format. General Instrument Corp. is the only company offering formant synthesis.

All of these semiconductor companies are producing memory devices that are appropriate for the storage of words or phrases and that can be played back through their integrated-circuit synthesizer chips on command. The naturalness and prosody (combination of pitch, duration, and intensity) of speech are preserved. Here is where the storage cost and flexibility of the systems vary. In the simplest systems, every phrase that the system will reproduce must be stored in memory and the phrases cannot be rearranged.

Constructive synthesis techniques based on analysis synthesis approaches are being used to overcome the twin handicaps of lack of flexibility and the high cost of semiconductor memory. These systems are all based on speech building-block approaches. Individual sounds derived from analysis synthesis techniques are joined together to create words and phrases. The individual sounds are based on speech elements called phonemes, or the extensions of these sounds, called allophones.

In one method, a directory of individual words can be stored in memory and, then, connected together to form phrases in a process called *word concatenation.* This approach is flexible because each word is stored only once and it can be used repeatedly to form many different phrases. However, this approach is rejected because of the unnatural and artificial quality of phrases formed from words strung together with inadequate coarticulation.

The maximum flexibility of synthesized speech is achieved with the largest number of elemental sounds. There are approximately 40 phonemes in the English language—14 to 16 vowel sounds and 24 consonant sounds. By contrast, there are 128 allophones based on phonemes and they incorporate the coarticulation variants of phonemes.

Virtually any word or phrase can be created with constructive synthesis techniques today. Moreover, the memory requirement is reduced to that which is necessary to store all the sound components. However, even the best of constructive synthesis techniques used today cannot match the quality of speech generated by the synthesis analysis methods. A constructive synthesis system with 256 elements in its speech library might need only 80 bits to produce the same 1-second utterance. This reduces the data storage requirement considerably—by factors of 100 or more. However, the quality of synthesized speech is degraded to the point where the quality of the speech and the natural inflection are lost. By contrast, most analysis synthesis systems use more that 1000 bits of data to produce one second of synthesized speech.

Constructive synthesis has an advantage over analysis synthesis because sequences of phonemes or allophones can be manually edited for immediate production and review. Analysis synthesis, by contrast, requires sophisticated data collection and analysis before synthetic speech is produced. Text-to-speech techniques for constructive synthesis considerably reduce the development time by rapidly converting a typed description of the words (to be spoken) into data sequences for immediate audition.

As various approaches to speech synthesis continue to be developed, some of their differences will be diminished. It is expected that competitively priced encoding systems for LPC will reduce the limitations of vocabulary development for analysis synthesis systems. Text-to-speech systems using constructive synthesis will also be available as low-cost integrated circuits.

A hybrid approach to speech synthesis that uses ICs to combine analysis synthesis and constructive synthesis methods in one system is another approach that has been developed. This system can provide an output that ranges from high-quality speech with a large data storage requirement to low-quality speech with unlimited vocabulary.

CHAPTER 2

SPEECH SYNTHESIS APPLICATIONS

Present and future applications of speech synthesis are seen as most significant in five activity areas: consumer products, industrial controls and production equipment, computer systems (including both word processing and data processing), telecommunications, and security systems. The availability of low-cost voice synthesis permits adaptations and innovations on these products that were not previously possible.

LEARNING AIDS

The first serious use of digital speech synthesis was in hand-held portable learning aids that were intended to amuse and entertain children while helping them to learn to read and spell. This concept has been enlarged with the introduction of language translators for both adults and children which do more than just replace the translation dictionary; they permit the user to hear the correct pronunciation of the foreign words and phrases.

While many of the consumer product applications for speech synthesis are subject to criticism as gimmicks or gadgets intended to increase cost and, therefore, profit for the manufacturer, they demonstrate that the low-cost system has arrived. For the average person, a clock that announces the time or a washing machine that announces the termination of a washing or rinsing cycle may seem to be novelties, but for the visually impaired they may make life just a bit more bearable.

INDUSTRIAL APPLICATIONS

The use of speech synthesis in industrial control rooms, power stations, nuclear plants, and other locations where a few people must monitor a myriad of controls can have many benefits. While these speech synthesizers would supplement rather than replace standard gauges, annunciator panels, or even alarms, they may easily pay for themselves in a short time by increasing the operator's ability to respond rapidly and correctly when a process has exceeded its normal limits. Appropriate phrases announced at the right time reduce the possibility of an operator making an incorrect response or wasting time analyzing the situation.

The use of vocal reminders may also be useful in the operation of certain kinds of industrial equipment where a failure to follow precise directions could lead to the destruction of property or even injury to personnel. These warnings could relate to an absent or inadequate flow of cooling oil, an improperly secured tool, or the wrong sequence of control settings.

USES IN MOBILE SYSTEMS

Vocal warning systems are now being used on automobiles to remind the driver of situations affecting his safety or the proper maintenance of the vehicle. Similar systems have a place in aircraft to supplement other warning indicators and readouts. In fact, several proposed aircraft collision-avoidance systems have synthesized speech as an integral element in their design. The pilot is actually directed to "climb," "dive," "make a sharp left turn," or told to follow other directions appropriate to the situation because he may not have the time to think out the best avoidance maneuver.

A recently introduced depth finder for yachtsmen and other small craft sailors reports the water depth as the ship maneuvers through the water. It calls out the depth (in feet) when the boat is in shallow water. This navigational aid can prevent the pilot from going aground when he is unable to make periodic depth readings because he is busy maneuvering around visible objects on the surface or through crowded harbors. The depth finder has a control that permits changing the reporting frequency of the vocal depth reports.

WIRING AND CABLING USES

Tape recorders have been used in the past as instructional aids by production personnel who perform nonroutine procedures like the fabrication of wiring harnesses or the wire-wrapping of electrical panels. Today, with the preparation of software, it is possible for a computer to scan wiring diagrams and generate verbal directions for carrying out wiring or cabling procedures. With available speech synthesis techniques, it is possible for the computer to dictate wiring procedures. By using constructive synthesis techniques, the operator can be told to "Connect blue wire from terminal position A-32 to terminal position J-24," etc. These procedures can be especially useful for telephone company or telecommunications systems installers who are called on to make extensive wiring changes in the field where the use of drawings or other instructional sheets are at best awkward and, sometimes, impossible. (The installer may be working on a tower or down in the confined space of an underground duct.)

WORD AND DATA PROCESSING APPLICATIONS

Synthetic speech is seen as having a place in such computer-related activities as word processing and data processing. Here it is employed in the on-going effort to make computers "friendlier." For example, it might be used to provide optional verbal instructions necessary in the performance of certain procedures. For example, if a word-processor operator is called on to make up a table or a statistical presentation, a voice synthesis routine could be employed to speed up the procedure and reduce the tedium of the operation. The voice could be used to prompt the operator in setting up columns, rows, and the values to be entered. Some people believe that operating aids or instructions should be discretionary—designed to be turned on or off by switch. They could be used to supplement instruction manuals in carrying out certain procedures.

Other applications appropriate for speech synthesis are seen in telecommunications. These include electronic filing and electronic mail where information is stored or transmitted in the form of speech. It might also be used in the area of secure voice transmission. The speech data is passed through a special encoder and, then, during reception later, it passes through a complementary decoder. Speech data encoded in this way cannot be deciphered without a knowledge of the encoding procedure.

INFORMATION AND SAFETY SYSTEMS

Speech synthesis could also have many different applications in fire and safety public-address systems, in addition to its use for public guidance, information, and direction. An electronic system could be devised with a long shelf life and with low-power requirements so as to respond after long periods of inactivity to emergency conditions—power failure or fire, for example. It could be used to make announcements about personnel evacuation—routes, doors, and other procedures. Systems of this type could be installed in public buildings, stores, or even in trains and planes. Similar systems, installed in elevators, could combine the announcement of floors (as an aid to the visually handicapped) with general directory information about the merchandise and services being offered on each floor.

AID FOR THE HANDICAPPED

Speech synthesis equipment and systems are rapidly becoming an important means of communication for the vocally impaired. Numerous systems have been devised to aid those deprived of a normal vocal tract due to a congenital defect, disease, or because of trauma. Until quite recently, these systems were bulky, inefficient, and awkward to use. Because of their high power requirements, they were not readily portable and, as a result, were usually fitted to wheel chairs.

A typewriter-type keyboard has been widely used as an input device but even this may be difficult or awkward for the individual who has other handicaps—poor eye/hand coordination or a paralysis of the arms and hands. Some systems now under development use "menus" of stock phrases in order to permit a rapid exchange of information when it is necessary for the user to gain emergency aid or meet pressing personal needs. The words and phrases also permit a rapid interchange of social protocol by conveying the user's interest in and need for a conversation. Some of the prototype equipment also includes an LED alphanumeric display to permit both the user and the listener to verify the synthesized speech. This is particularly useful if the handicapped person is also deaf and, therefore, unable to listen to and determine the accuracy of the words or phrases being announced. Mistakes in keying could prove puzzling and embarrassing to both parties.

Psychiatrists and psychologists agree that the inability to speak may be more debilitating to a person than the loss of sight or hearing because it leads to the total isolation of the person afflicted. An operation on the larynx to permit rudimentary speech by a person who has had his vocal cords removed surgically (as a cancer cure) has proven successful in many cases. But, the patient must learn a new breathing and speaking process that is slow and laborious. A speech synthesizer could function as an interim aid or, perhaps, could even preclude the need for such retraining.

Engineers working on prosthetic speech-generation equipment are most challenged by the interfacing problem for the speech disabled who are partly paralyzed and, thus, unable to employ a keyboard or keypad. Some experimenters are working on procedures that permit the selection of words or phrases by the tracking of eye movements. An electro-optical system, focused on the eye, translates eye movements into two-dimensional coordinates and permits the selection of specific entries on a "menu." This permits "eye pointing" that is analogous to "finger pointing."

CHAPTER 3

HOW IS HUMAN SPEECH PRODUCED?

Studies of speech patterns have shown that the spoken language can be analyzed into simple sounds that can be distinguished by the human ear. The "stringing" together of these basic sounds, called *phonemes*, permits the reconstruction of words, phrases, and sentences. However, care must be taken in the assembly or concatenation process to assure natural sounding speech. This calls for a means of reintroducing speaker inflection, volume, and emphasis—all necessary speech attributes for rapid human comprehension.

PHONEMES

American English has about 40 phonemes—16 are vowel sounds and 24 are consonant sounds. Phonemes are further classed as *voiced sound*, as in "eye," or *unvoiced sound*, like the sh in "shy." Experiments have shown that there are fewer unvoiced sounds than voiced sounds and they depend less on the inherent characteristics of the speaker than do voiced sounds. This has become a very important fact in electronic speech synthesis and it has influenced the direction of research.

In normal speech, from 10 to 15 phonemes are spoken every second. With an effective method for classifying and storing phonemes in memory, it is a relatively simple matter to call out and "string" together the appropriate phonemes needed to form words, phrases, and even complete sentences. Since the 40 phonemes can be coded with a 6-bit code, speech can be generated at a rate of 60 to 90 bits per second. The catch to this method of coding phonemes is the omission of the many important speech attributes. Solving this problem with minimal expenditure of data is the central thrust of current development. As a first step in one approach, the number of phonemes has been expanded to account for all speech variations and a new set of 128 *allophones* has been derived. An allophone is a variation of a phoneme as conditioned by its position or by adjoining sounds.

HUMAN VOCAL SYSTEM

An understanding of human voice production is useful in any discussion of electronic speech synthesis because it gives insights into the problems which must be overcome in the perfection of low-cost effective systems. The human vocal system consists of the vocal tract and the nasal tract. The vocal tract is an acoustic tube of variable cross-sectional area that extends from the vocal cords to the lips. The nasal tract is a secondary cavity that is coupled to the vocal tract by the flapper action of the velum.

Resonant Chambers

Humans are able to speak because of the interaction between the lungs, the vocal cords, and the vocal cavity. The lungs are a source of power for producing speech; they force air past the vocal cords and through the vocal cavity. The vocal cavity is made up of the lips, the tongue, the teeth, and the jaw. Fig. 3-1 is a simplified diagram of the vocal system. By controlling the resonant chambers of the throat, mouth, and nasal cavities with the mouth position, the tongue position, and the throat orifice size, a speaker can generate a phoneme. The lungs and the vocal cords are able to produce *impulses*, termed voiced sounds, and *hisses*, classed as unvoiced sounds.

The human vocal tract is a flexible organ pipe whose cross-sectional area may be varied voluntarily anywhere along its length from the vocal cords to the lips. A sound tube about six inches long (the size of a toothpaste tube), the vocal tract's cross-sectional area may be varied at the throat end from a complete closure to a full opening of about three square inches. This variance is caused by the positioning of the lips, jaw, tongue, and soft palate. The *soft palate* (or *velum*) at the back of the

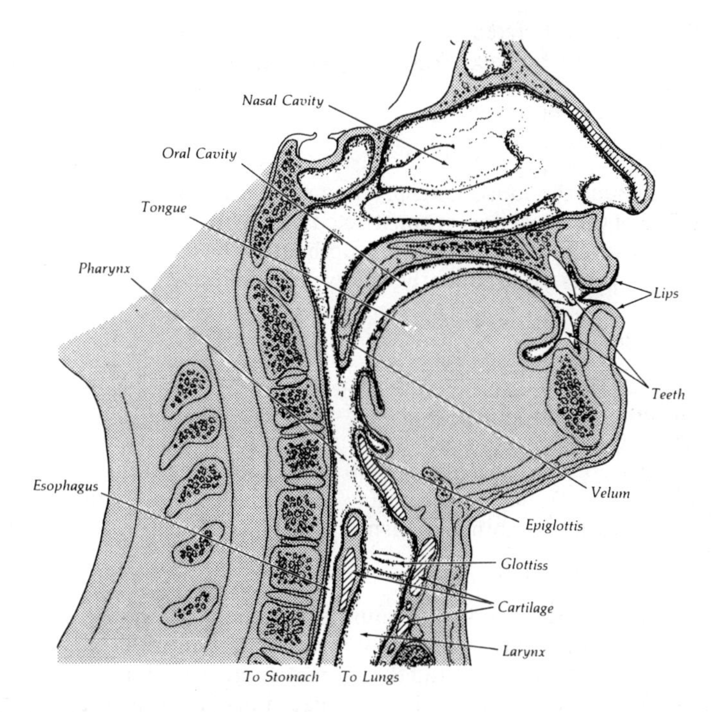

Fig. 3-1. The human vocal system.

roof of the mouth is flexible and acts as a flapper valve to connect the nasal tract to the vocal tract. (Important in speech formation, the nasal tract is a cavity about five inches long which is able to act in parallel with the vocal tract.)

The human vocal tract produces two different kinds of sounds: periodic voiced sounds and aperiodic unvoiced sounds. Voiced sounds are the vowel sounds that are produced by forcing air from the lungs through the slot between the vocal cords. Airflow through this passage, called the *glottis*, vibrates the vocal cords which, in turn, modulate the airflow. A range of sounds is possible because the glottal orifice can also be opened to an area of about three square inches.

Fricative Sounds

Unvoiced sounds are further divided into *fricative* and *plosive* sounds with reference to their physiological origin. Fricative sounds are the consonants, which include s, sh, f, and th. They are generated when the vocal tract is constricted at some location along its length and are created by the turbulence produced when air is forced through that constriction.

Plosive Sounds

Plosives include the consonants p, t, and k. They are produced when the throat end of the vocal tract is constricted by the lips or the tongue. They are formed when air pressure, built up behind the closure, is abruptly released. Plosives are frequently followed by fricative sounds.

Voice Frequencies

The vocal system may be considered as a time-varying filter that is able to condition the sound waves that are generated by a broad range of sources. The spectrum of voice frequencies extends from about 100 Hz (cycles per second) to more than 3000 Hz. It should be pointed out that the voiced and voiceless sources may be combined. For example, two sound sources are combined in forming the voiced fricative consonants v and z.

The interaction between sound sources and the vocal system is loose enough so that they may be represented as separate entities—very much the same as a vibrating reed and the horn of a clarinet. Thus, the sounds from the mouth are related both to the spectrum of the source and to the physical characteristics of the combined vocal and nasal tracts.

ORGAN-PIPE THEORY

Basic organ-pipe theory can be applied to the vocal system. The harmonics from the periodic voiced sounds are inversely related to the square of the frequency. There are favored frequencies or "poles" that correspond with the acoustic resonances of the vocal tract. These resonances are called *formants*. The wavelengths of speech turn out to be comparable to the length of the vocal tract. The vocal tract can be considered as a kind of horn that is open at the mouth end and is closed at the vocal cord end. The 6-inch length of vocal tract approximates the odd quarter wavelengths of its resonant frequencies.

For vowel sounds in which the nasal tract is closed off, the resonant frequencies are approximately 500, 1500, and 2500 Hz. When the nasal tract is included in the system, another resonant frequency of about 1000 Hz is introduced. Voiced sounds are produced by changing the shape of the vocal tract which, in turn, changes the resonance frequencies. Fig. 3-2 shows a typical frequency spectrum for voiced speech. By contrast, unvoiced sounds are derived from a broad uniform noise spectrum. However, the constriction employed to form the sound is positioned at some location along the tract. This serves to change the length of the resonant system and shifts the frequencies of the resonances. Again the voice is affected by the frequencies that are favored or suppressed by the resonant system.

As a result, the various formant resonances move back and forth as the shape of the vocal tract is altered. However, the formants change their frequencies slowly with respect to the rate of pressure fluctuations in the speech wave. This can be seen in a sound spectrogram. A spectrogram shows how sound energy is distributed at various formants. Fig. 3-3 is an example of a spectrogram.

Over the past 200 years, a number of attempts have been made to build speaking machines by duplicating the physiology of the human vocal tract. The success of these experiments was limited both by a lack of knowledge of that physiology and a lack of suitable materials for modeling the tract. Historical records indicate that Christian Gottlieb Kratzenstein won a prize offered in 1779 by the Imperial Academy of St. Petersburg in Russia. The academy wanted to find out how vowels are formed and thought the answer could be found by building a simulator that employed organ-pipe theory as it was understood at the time.

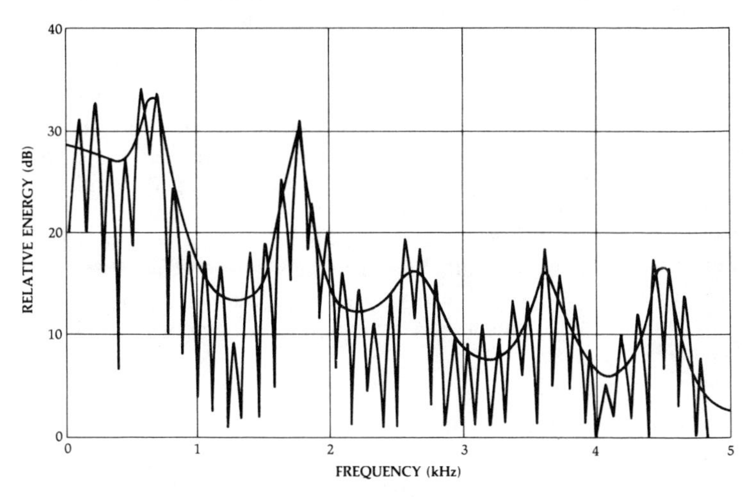

Fig. 3-2. Typical frequency spectrum for voiced speech.

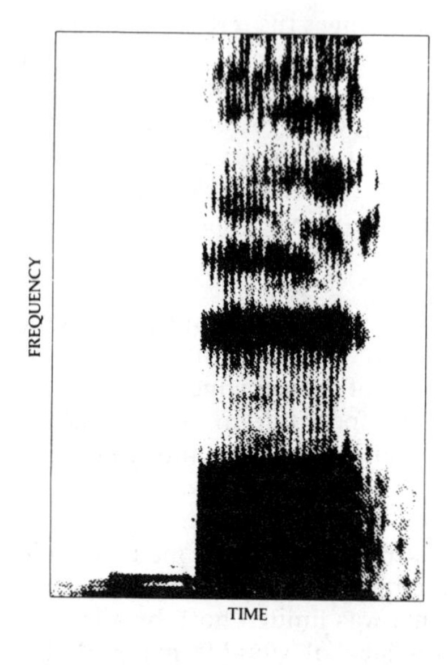

Fig. 3-3. Voiceprint of the sound "ah." The X axis is time; the Y axis is frequency. Amplitude is displayed as a function of print density.

Kratzenstein, limited by the availability of suitable flexible materials, won the prize by constructing a set of five resonators, each of which represented the form and dimension of the human vocal tract as it formed one of the five vowel sounds, A, E, I, O, and U. The five resonators were driven with vibrating reeds that simulated the vocal cords by periodically modulating the air stream.

In about 1792, an Austrian scientist named Wolfgang von Kempelen built a machine that not only reproduced vowel sounds and consonants better than Kratzenstein's resonators, but it could also make speech-like sounds. Although von Kempelen's machine was far more sophisticated than the resonators, it was not taken seriously because its inventor had earlier lost his credibility by perpetrating a hoax on his scientific colleagues. A "chess-playing machine" that he had built was revealed to be occupied by a legless Polish war veteran who had concealed himself in a small compartment underneath the machine and who was making each of the chess moves.

Von Kempelen's mechanical speaking machine was a strange-looking contraption that consisted of a bellows, an air chamber, pipes, a reed, levers, and whistles. It is illustrated in Fig. 3-4. The operator used one hand to manipulate a kind of leather tube that acted as a resonator while

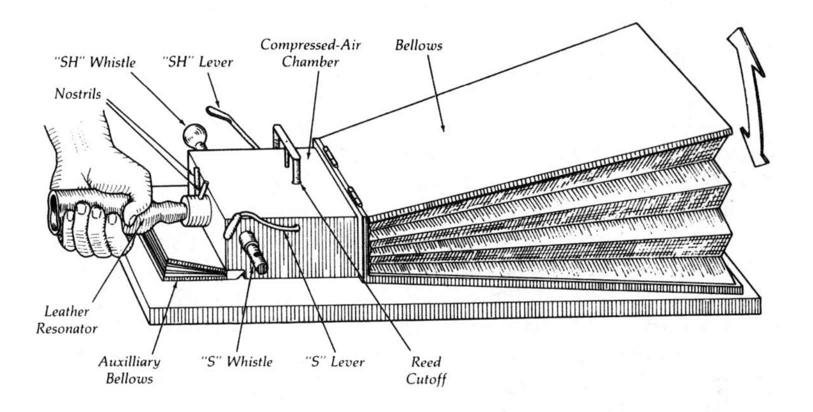

Fig. 3-4. Mechanical speaking machine devised by von Kempelen in 1792.

his other hand controlled four separate constricted passages that were used to simulate consonants. It was, in effect, a kind of bagpipe.

Sir Charles Wheatstone, a musical instrument maker and an electrical scientist, built a copy of the von Kempelen machine which he placed on exhibit in England. Wheatstone, whose invention of the Wheatstone bridge assured him a place in the history of electrical engineering, influenced yet another famous inventor—Alexander Graham Bell. Bell was so impressed by the speaking-machine reproduction that he undertook to build one of his own. This version went a lot further in that it actually used suitable materials to represent the rigid and flexible elements of the voice tract. The human physiology was modeled with flexible plastic materials, such as gutta percha, wire, and rubber supported by wood and metal. It was reported that the machine produced both vowels and consonants.

One of the more successful precomputer era speech synthesizers was the *electrical Voder*—short for voice-operation demonstrator. It was first exhibited at the New York World's Fair in 1939. The Voder employed two sound sources: a wideband noise source and a periodic oscillator. The outputs of these noise sources were altered as they passed through a resonance control chamber. Analogous to the vocal tract, this resonance control chamber contained ten contiguous bandpass filters that covered the range of normal speech frequencies. The outputs of the bandpass filters were modulated keys that controlled the gain. Additional keys provided pulse excitation of other filters to simulate three plosive sounds: t–d, p–b, and k–g. A wrist bar permitted the selection of either the random noise source or a relaxation oscillator and a foot pedal controlled the pitch of the oscillator.

At the time that the Voder was being demonstrated, another system called a *Vocoder* was invented to investigate methods for reducing the bandwidth of transmitted speech. The Vocoder took periodic samples of the amplitude spectrum of a speaker's voice at a number of different frequencies. From this information, it derived electrical analogies of the relatively slow changes in the samples. A voltage was also derived to represent the voice pitch and to indicate whether the sound was voiced or unvoiced. Since only these slowly changing quantities or parameters were transmitted, they required less bandwidth than a normal voice channel. At the receiving end, the transmitted quantities were used to resysnthesize the original speaker's voice.

In work performed at the Bell Laboratories on speech synthesis in the 1960s, two scientists, John L. Kelly and Carol Lochbaum, developed a right circular cylinder analog model of the human vocal tract. It was the basis for studies that led to the development of an electrical filter that could simulate the vocal tract. Bell Laboratories' interest in speech synthesis was directly related to their research into speech compression and other means for reducing the total amount of data needed to transmit an intelligible human voice.

The Kelly-Lochbaum analogy consisted of a string of 10 tubes. Each tube was 17 centimeters (about 7 inches) long. The diameters of each of the tubes in the 6-foot long series were independently variable. Sound energy entering one end of this variable diameter duct would pass down its entire length. Ten sections were selected as a reasonable number necessary to represent the virtually infinite number of diameter changes possible in the human vocal tract. The concept is illustrated in Fig. 3-5.

The mechanical analogy was developed so that scientists could study the formation of forward and reverse waves and the generation of harmonics at each interface. The effect obtainable on the output was the cumulative result of these effects at each of the nine transitions. The experimental setup could be viewed as a large segmented flute or even as an audio analogy of a microwave waveguide with nine different transitions in the interior guide dimensions. The effects were directly related to the differences in the areas of the adjacent sections and the direction of propagation.

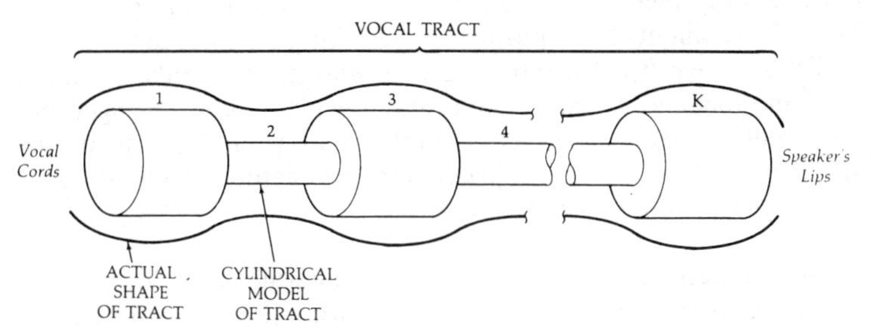

Fig. 3-5. Physical analogy of the vocal tract as used in linear predictive coding.

As a result of these experiments, scientists were able to derive the electrical Kelly-Lochbaum filter. The Bell Laboratories scientists found they could simulate the effects of sound at the input to each stage mathematically and could account for all the reflection coefficients with four multiplication and two addition steps. However, a complete calculation called for a stupendous amount of calculation. The results of each stage, in succession, were needed to calculate the next stage and this meant that a time delay was involved. It was not possible to do all the calculations simultaneously. For all practical purposes, only a large-scale mainframe computer was capable of doing these calculations.

Two scientists from Nippon Telephone and Telegraph, Fumitada Itakura and S. Saito, who were doing research work at Bell Laboratories, studied the Kelly-Lochbaum filter and devised a method for reducing the number of calculations. They applied a method known as *least squares estimation* and found they could make drastic reductions in the amount of calculation needed. The method was based on *linear prediction,* a technique devised by Karl Gauss (of magnetism fame) back in the 1790s. It permitted the prediction of events in a linear system using information derived from previous samples.

In 1965, when Itakura and Saito applied their linear predictive coding to the filter, they found that they could combine terms mathematically and eliminate two of the multiplication steps. However, even after this simplification step, some 400,000 calculations (200,000 additions and 200,000 multiplications) were required every second just to simulate the vocal tract although it actually changes shape very slowly. It was also necessary to consider the time delay that related to the cumulative nature of the calculations. This placed an additional burden on even the highest-speed computers.

Approaching speech synthesis from an entirely different point of view, Texas Instruments Incorporated was looking for ways to incorporate a realistic voice output into toys, games, and learning aids. Of necessity, an economical, low-cost, reliable, and low power-consuming approach was needed. Texas Instruments sought to carry out this objective by the application of digital microprocessor and memory techniques which they understood and knew how to use.

Although many other individuals and companies were looking at various ways to carry out speech synthesis within the more limited context of microprocessor technology, TI was the first company to introduce a working system in a commercial product. TI got around the complexities of the two-stage filter by developing and implementing (on a silicon chip) a filter containing an array multiplier, an adder, some memory, shift latches, and internal feedback. The single-stage filter with a single multiplier and an adder then became the equivalent of a 10-stage filter with two multipliers and two adders per stage.

Because it takes four times as long to perform digital multiplication as it does addition, it was necessary to work out a delay period so that the addition and multiplication processes would be done simultaneously. This was accomplished by means of a pipeline multiplier. Multiplication was, in effect, given a head start and, by performing four multiplication steps simultaneously for four clock periods, the addition and the multiplications steps would coincide at the end of the fifth time period. Thereafter, a product and sum would be available at each clock beat.

CHAPTER 4

TIME DOMAIN ANALYSIS SYNTHESIS

In Chapter 1, the general concepts of the two most widely used techniques for electronic speech synthesis, time domain synthesis and frequency domain synthesis, were described. Time domain synthesis is based on the formation of a synthetic waveform that is interpreted by the human ear as if it were the original speech waveform. It completely ignores the techniques of voice generation but it is concerned with the physiology of human hearing.

The goal of time domain synthesis is not exclusively the reduction of the number of bits required for conventional analog-to-digital conversion by pulse code modulation. It is actually concerned with the extraction of sufficient information from the original time domain waveform to permit quality synthesis of that sound. After this is accomplished, time domain synthesis seeks to remove the remaining redundant information from the waveform so that the waveform can be represented with the least number of bits.

Studies have shown that most of the audio quality and recognition factors of human speech are contained in the frequency spectrum of the signal. It has been found that the human ear is most sensitive to the power spectrum of the speech signal. Sensitivity to the phase relation of the component frequencies is secondary. Both time and frequency domain synthesis are based on these observations. Thus, in frequency domain techniques, phase information is ignored and the signal power spectrum is reproduced as accurately as possible with a limited number of frequency variables.

DELTA MODULATION

One of the earliest methods employed to compress speech data was a time domain coding technique called *delta modulation*. This method is based on the assumption that the analog signal is always either increasing or decreasing in amplitude. The typical sampling rate is 64,000 times per second. Each sample is then compared to the estimated amplitude value of the previous sample. If the first value is greater than the estimated value of the latter, the slope of the signal generated by the model is positive. If not, the slope is negative. The magnitude of the slope is chosen so that it is at least as large as the maximum expected slope of the input waveform signal.

Another form of delta modulation is known as *continuous variable slope delta modulation (CVDS)*. It allows the slope of the generated signal to vary. This technique permits a data sampling rate of 16,000 to 32,000 bits per second, or approximately half the 64,000-bit rate of standard delta modulation.

After sampling the original waveform, the output closely resembles the original waveform. Then, the phases of the component frequencies are chosen so that the resulting time domain waveform can be represented by a small number of bits. Thus, while the frequency domain technique ignores phase information, time domain synthesis uses the phases to find a waveform whose representation requires the least number of bits.

MOZER'S TECHNIQUE

Another time domain analysis encoding technique takes advantage of the periodic repetitions in the human speech waveform. Developed by Forest Mozer, it relies on the fact that the human ear cannot detect a shift in the phase of a voice waveform. This permits a certain rearrangement of the original waveform and it permits a further reduction in the amount of memory required to store the speech. Mozer uses a number of steps to compress the information in the waveform.

1. Phase-angle adjustment to obtain a waveform with a pitch period that is symmetrical with respect to time. This permits the redundant or mirror-image half to be deleted.
2. Half-period zeroing; a method that replaces the low-level portion of the pitch period with silence.
3. Digital compression using delta modulation to minimize memory requirements.
4. Elimination of redundant or similar speech segments by the repetition of pitch periods.

The operation of waveform analysis will be explained later in Chapter 8 with the description of a speech synthesis device that is now available from National Semiconductor Corp.

THE SPEECH WAVEFORM

In time domain synthesis, the speech waveform is broken up into segments called periods. The natural period for voiced waveforms is the period of vocal cord oscillation. However, there is no natural period for unvoiced waveforms so a fixed number of input samples must be selected.

The first step in time domain analysis is to divide each period into a small fixed number of samples. This number may be less than half the original number of samples per period. It is reasonable to select a fixed number of data samples per pitch period because the total amount of information in each period can be expected to be about the same, independent of its duration. Selection of pitch period with a fixed number of samples is determined by consideration of the power spectrum of the new waveform; it should approximate the original waveform's spectrum as closely as possible.

This decrease in samples per period is the first reduction in data rate. It is achieved without a loss of speech quality because the original 25.6-kHz sampling frequency is far higher than what is needed to reproduce the 5-kHz frequencies that are at the upper limit of human speech.

The number of samples chosen per period in National Semiconductor's first integrated-circuit time domain synthesizer was 128. This will be reduced to 64 samples per period in an improved version that the company has said it is ready to introduce. The decreased samples per period halves the bit rate because the number of bits per sample required by the two synthesizers is the same. Reduction in the number of samples, coupled with other improvements, has actually increased the frequency response and speech quality.

In addition to matching the power spectrum of the new waveform as closely as possible to the original waveform—another of the criteria for fixing the number of samples per period at a smaller number—the duration of each pitch period must be matched as closely as possible to the original period length. If the sampled information were to be played back at the original 25.6-kHz sampling frequency (or at any frequency for that matter), the duration of the pitch periods would not be correct. The original variable-length pitch periods would now all have the same duration.

In the original National Semiconductor time domain synthesizer, this problem was overcome by having 32 different playback frequencies. The output frequency was chosen for each period so as to give the closest approximation to the original period length. In the latest National Semiconductor time domain synthesizer, the number of samples of output data per period is fixed at 64, and the playback frequency is fixed. However, a variable number of silent samples is added at the beginning of the pitch period to control pitch-period duration. Thus, instead of storing a playback frequency for each period, the number of silent samples that are to be added to the data is stored.

PITCH AND PHASE

Pitch is changed in the new National Semiconductor synthesizer by adding or subtracting silent samples to the period. It is possible to obtain a new intonation by changing the stored length of silence for each period rather than changing the output data. Using a set of fixed length periods and an associated table of silent lengths for adjusting the pitch of each period, the next step in time domain synthesis is the reduction of data required for each period. This calls for scaling the amplitude of each pitch period. Experiments at National Semiconductor Corp. have shown that a 3-bit sampling (to set the overall amplitude for the entire period) results in a 25% reduction in data rate over a 12-bit sampling, with little degradation in speech quality.

After amplitude scaling is performed, the phase is adjusted. A discrete Fourier transform is performed on each pitch period to provide a set of amplitudes and phases for the component frequencies in each period. In the latest devices from National Semiconductor, there are only 32 sets of each—half as many as in the earlier models.

Amplitudes of component frequencies correlate closely with the quality of speech because they represent the input power spectrum. The ear responds primarily to this power spectrum. The 32 phases may be varied to find a waveform that can be represented by less than sixty-four 12-bit samples per period. The power spectrum will be correct because the amplitudes of all frequencies are fixed and only the phases are varied.

DATA COMPRESSION

The first step in data compression is the selection of phases so that the time domain waveforms are symmetric within each period. With a symmetric waveform for each component frequency, the overall waveform will be symmetric. This means a reduction in data samples is possible because the second half of each period will be the mirror image of the first half. This is illustrated in Fig. 4-1. The top waveform shows a period of speech while the bottom waveform shows the change when one possible set of symmetric phases is chosen.

It should be noted that there is much less power in the beginning and ending quarters of the pitch period than there is in the two middle quarters where most of the power (largest amplitude) is concentrated. As a result, the two outside quarters of the waveform can be discarded because the energy levels are too low to be heard. Remember, also, that the ear cannot detect phase differences in the power spectrum. Therefore, both the top and bottom waveforms in Fig. 4-1 will sound exactly the same although they look different because of the phase shift. As discussed earlier, a 2 to 1 data reduction, or compression, is made possible because of this characteristic of the ear. The parts of the waveform that are discarded are replaced by a constant silence level which does not require any memory in the synthesizer chip or ancillary memory.

Further compression is afforded by a close representation of the actual amplitude of each speech sample. The delta modulation technique employed permits the storage of just the changes in amplitude rather than the absolute amplitude. This reduces the amount of data to be stored per sample to 2 bits as compared with the 12 bits of the original signal.

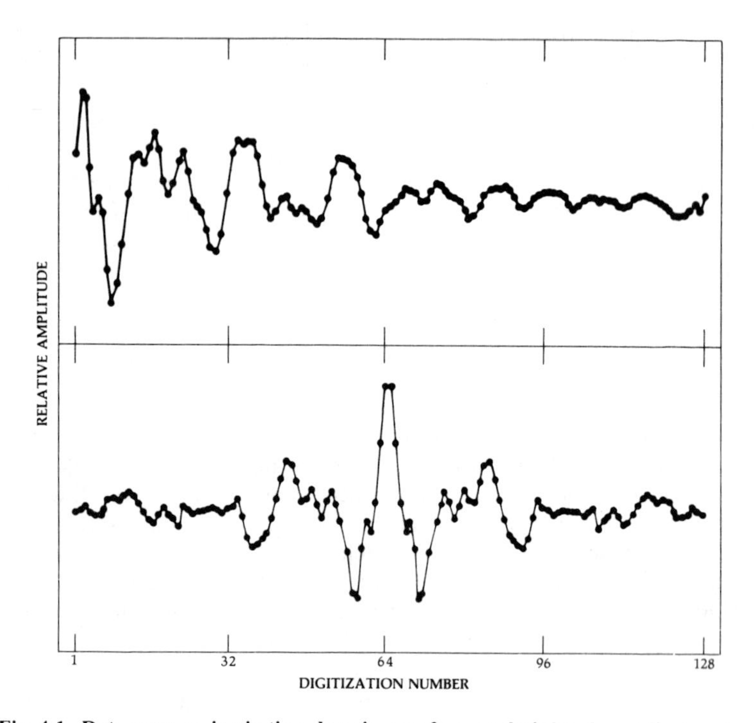

Fig. 4-1. Data compression in time domain waveform analysis by phase adjustment.

The final step in the compression scheme is to repeat a single period several times in order to replace similar periods. This further cuts the amount of speech-pattern data that must be stored. The same pattern may be repeated three or four times with voiced waveforms and as many as seven or eight times with unvoiced waveforms without any noticeable loss of quality.

The data compression techniques used by the National Semiconductor *DIGITALKER*™ system that is now in production will permit the quality reproduction of a word (with a male English-speaking voice) using about 1000 bits of data stored in memory. Words spoken by women and children call for about 25% more data, or about 1250 bits, to be stored in memory because of their higher pitch. However, National Semiconductor Corp. claims that substantial improvements in data rate, which have been demonstrated in the laboratory, may help to reduce this difference.

*DIGITALKER is a trademark of National Semiconductor Corp.

CHAPTER 5

FREQUENCY DOMAIN ANALYSIS SYNTHESIS

Historically, the most widely followed approach to speech synthesis has been by means of frequency domain synthesis. All of the simulation experiments of Kratzenstein, von Kempelen, Wheatstone, and Bell were attempts to model the human voice by combining noise sources representing voiced and unvoiced vocal tract excitations coupled with output filters. A serial array of multiple filters represented the different resonant conditions possible within the vocal tract.

In later work, made possible with the use of the digital computer, digital filters were substituted for resonators and various digital signal generators replaced the analog noise sources. When the experiments progressed to the point where the waveforms could be stored as digital data, frequency synthesis techniques stored representations of the source excitations and the parameters of the output filters that these excitations drive.

By storing only a small set of parameters that vary slowly with time, frequency domain synthesis achieves a significant bit rate reduction. The currently available integrated-circuit frequency domain synthesizers switch between on-chip source excitations and output filter states. They reproduce a facsimile of the time-varying frequency spectrum of the original speech waveform by using the control data and filter parameters stored in digital memory.

LINEAR PREDICTIVE CODING

There are a number of different approaches within the general classification of frequency synthesis domain that include linear predictive coding (LPC) and formant coding. One of the most important analysis synthesis concepts, linear predictive coding, or LPC, is based on the modeling of the vocal tract as shown in Fig. 5-1. There are two noise sources: (1) a Periodic Impulses source for voiced sounds, and (2) a White Noise (or hiss noise) source for unvoiced sounds. By an appropriate selection of these sources with a single-pole double-throw switch (permitting the voiced/unvoiced decision), the equivalent of an excitation signal is formed. While pitch may be adjusted at the voiced sound source, the amplitude of the excitation signal is adjusted prior to its application to the vocal tract model.

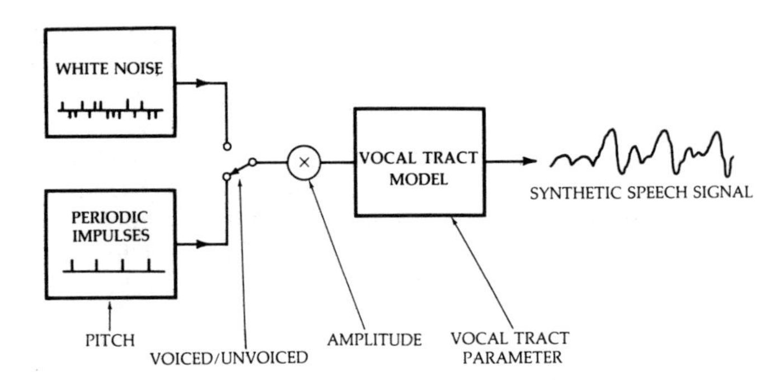

Fig. 5-1. Elements in a speech synthesis model.

The analogy of the vocal tract is a filter that acts as a modulator of the excitation signal. The analogy of resonances is derived by varying the characteristics of the filter with reference to time. This produces specific frequency bands or *formants*. The greater the amount of control that can be exercised over the filter, the more accurate is the approximation of speech.

Fig. 5-1 can be used to represent either an analog version of the vocal tract model or a digital version. In the analog version, both voiced and unvoiced sources would be audible. However, in the digital version, the output of both sources would be a stream of digital data and the filter would be designed to operate on digital data. The digital linear predictive coding speech synthesizer can produce a constant synthetic approximation of speech when energy, the voiced/unvoiced decision, pitch, and amplitude have been specified. However, the output must be converted to an analog form.

LPC DATA PREPARATION

In the analysis synthesis linear predictive coding (LPC) method for generating electronic speech, the voice is recorded, digitized, and computer analyzed to extract the energy, pitch, and vocal tract variables needed to reconstruct the original speech. After the speech data is analyzed by the computer, it is coded in a format called the *frame repeat format* (FR) for use with a specific synthesizer. At this point, the speech can be auditioned and manually edited to improve its quality and/or reduce the data rate. The final step is to format the edited data for storage in a voice-synthesis memory or some other data store. The memory becomes the source for speech data to the synthesizer.

The amount of speech data that can be stored in one memory device depends on how the data is edited and, of course, the capacity of the memory. After editing for maximum quality, the data rate is reduced to approximately 1500 to 1700 bits per second. Further editing can then be performed to reduce the data rate as low as 1000 bits per second but this may degrade the speech quality.

Synthesis of an average word by the LPC method requires about 640 bits of speech data. Thus, a 128K-bit ROM can store data that is equivalent to approximately 200 words of speech.

A COMMERCIAL SYSTEM

The first commercially successful three-chip system for speech systhesis was developed by Texas Instruments for use in its learning aid for spelling called *Speak and Spell*™. The set consisted of a speech synthesizer (TMC 0280), a read-only memory (TMC 0350), and a controller (TMC 0270). The system is based on the voice-compression technique called linear predictive coding (LPC). The merit of LPC is in its ability to generate high-quality speech from data rates of less than 2400 bits per second. This can be done because the synthesizer chip contains a pipeline multiplier, and adder/subtracter, and delay circuits to do the work of 10 filter stages.

To produce speech, the controller gives the synthesizer the starting point of a string of data stored in the ROM. The 128K-bit (actually 131,072-bit) ROM can store about 165 words or 115 seconds of speech depending on data rate, which may range from about 600 to 2400 bits per second. The ROM provides the pitch, amplitude, and filter varibles from which the synthesizer chip constructs the speech waveform.

To update the speech variables, the synthesizer must transfer a new parameter string from the ROM every 20 milliseconds. However, it turns out that a complete set of parameters is not always required because of redundancies in a typical speech pattern. This whole section of the data stream may be replaced by a single repeat bit, reducing it from a maximum of 49 bits down to a minimum of 4 bits and, thus, saving ROM space. However, LPC is the process that really reduces the data to storeable proportions.

A VOICE SYNTHESIZER SYSTEM

LPC uses a digital filter to model human speech. Input to the filter is either a periodic or random sequence of pulses and its output resembles the speech waveform. The process is analogous to the actual speech physiology, in which air is forced past the vocal cords by the lungs and the rest of the vocal tract modifies the sound.

Fig. 5-2 is a block diagram of the basic elements of an LPC voice synthesizer system. The vocal tract is modeled by a multistage lattice filter that uses coefficients k_1 to k_n to filter digitally an amplified excitation signal. The filter output is then passed to a digital-to-analog converter connected to a speaker.

The excitation signal may be either a periodic sequence of pulses or pseudorandom noise. Periodic sequences generate voiced sounds (typically vowels) which are created by the vibration of the vocal cords. The rate at which the vocal cords vibrate determines the pitch of the sound generated. Pseudorandom white noise generates unvoiced sounds (fricative and plosive) which are heard when the vocal cords are held open and the vocal tract is constricted at some location along its length.

The coded digital data necessary to specify the type of excitation, degree of amplification, and filter coefficients are stored in ROM. Pitch bits

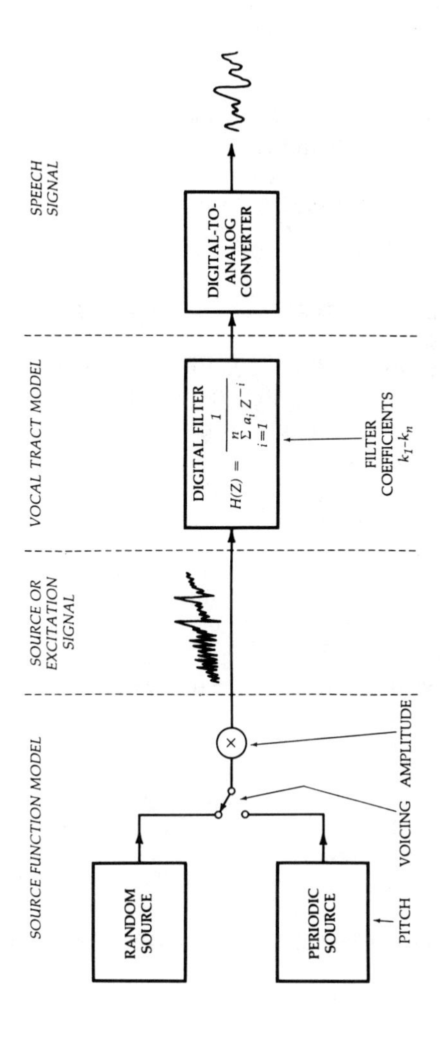

Fig. 5-2. A basic speech synthesis model showing the control parameters for linear predictive coding.

either vary the frequency of the periodic sequence or, if all are zero, select random data as the lattice filter's excitation. In addition, a multibit amplification factor is stored. It adjusts the constant-amplitude signal from the excitation source to produce sounds of varying intensity.

The filter coefficients are typically updated every 20 milliseconds. This results in quality speech and reasonable ROM storage requirements. The lattice filter will model the vocal tract dynamics even more closely if the update rate is increased, but this will increase the amount of data that must be stored in ROM.

A SPEECH SYNTHESIZER

The lattice filter of the TMC 0280 speech synthesizer has 10 stages (called 10th-order LPC or LPC-10). The first 9 stages carry out two multiplications and two additions on their two digital inputs before passing the results backwards and forwards to their neighbors. The operations of the 10 stages are carried out sequentially, as are the 4 operations within each stage.

With an appropriate timing of this sequence of 40 operations, it is possible for one adder and one multiplier to do the work of twenty adders and twenty multipliers. However, it is easier to understand how this timing works by visualizing 10 stages carrying out these 40 operations—one after another. The block diagrams shown in Figs. 5-3 and 5-4 model the occurrence of actual additions and multiplications with hypothetical adders and multipliers.

Fig. 5-3 shows the excitation signal **u** entering Stage 10 of the filter and the output from Stage 1 being applied to the d/a converter and, also, being fed back through the filter. (Only 3 stages are shown in detail since the other seven are just like Stage 1.) Stage 10, with only a single output, has no need for the second adder or its associated multiplier. Nevertheless, the multiplier in that stage is used as an amplifier.

Fig. 5-4 illustrates the filter's **n**th stage during the **i**th time cycle in response to the **Y** data derived directly from **u** and the **b** feedback data. In this presentation, the subscript of the **Y** and **b** data defines the stage in which that data are used while the term in parentheses indicates the time cycle in which the data are generated.

In all of the calculations performed by the filter, the **Y** and **b** data, as

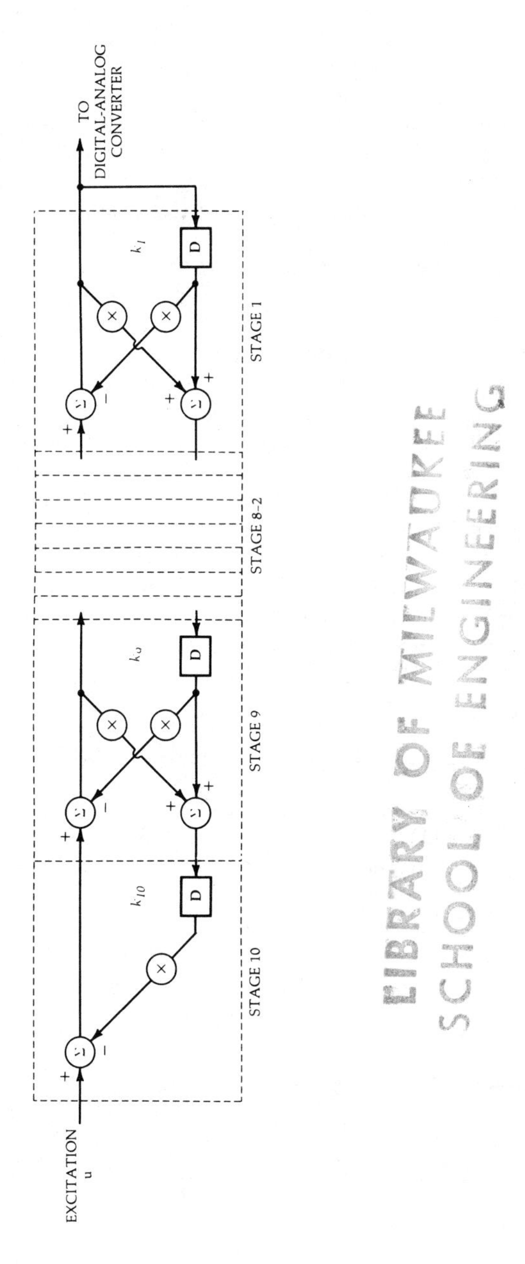

Fig. 5-3. A hypothetical 10-stage lattice filter.

well as the coefficients k_1–k_{10}, are multibit numbers. The coefficients k_1–k_{10} may vary between +1 and -1 and are periodically updated.

A block diagram of the actual digital 10-stage lattice filter is shown in Fig. 5-5. The filter includes a pipeline multiplier, an adder/subtracter circuit, a delay circuit, a shift register, and a latch memory. The pipeline multiplier output is one input to the adder/subtracter.

The pipeline multiplier performs all twenty of the multiplications required by the lattice filter. It receives either Y_n or b_n data via the data bus, and the coefficients k_1–k_{10} from the k stack of the shift registers by the data lines. It initiates a different multiplication operation at every 5-microsecond interval. One multiplication operation requires 8 time periods, so the pipeline multiplier has 8 stages. Therefore, there are 8 multiplications that are invarious stages of completion at any given time.

The coded data are applied to the digital 10-stage lattice every 20 milliseconds. A 10-kHz output speech sample rate requires a time cycle of 100 microseconds. This allows the basic operations of the adder/subtracter, the multiplier, and the shift registers to be accomplished in 5-microsecond time periods. Because these speeds are well within the capabilities of p-channel MOS technology, PMOS technology was used to make the TMC 0280 synthesizer chip.

THE TMC 0350 ROM

The coded speech data for the synthesizer chip is stored in the TMC 0350 ROM, organized as 16384-by-8 bits. The ROM has an internal 18-bit address counter/register and two 8-bit output buffers. Fourteen

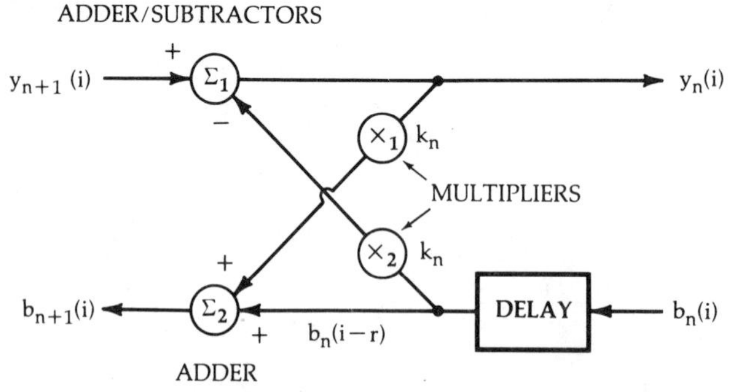

Fig. 5-4. One stage of lattice filter given in Fig. 5-3 is shown in detail.

bits of the address go directly to the ROM, while the four most significant bits are used in a 1-of-16 Chip Select function.

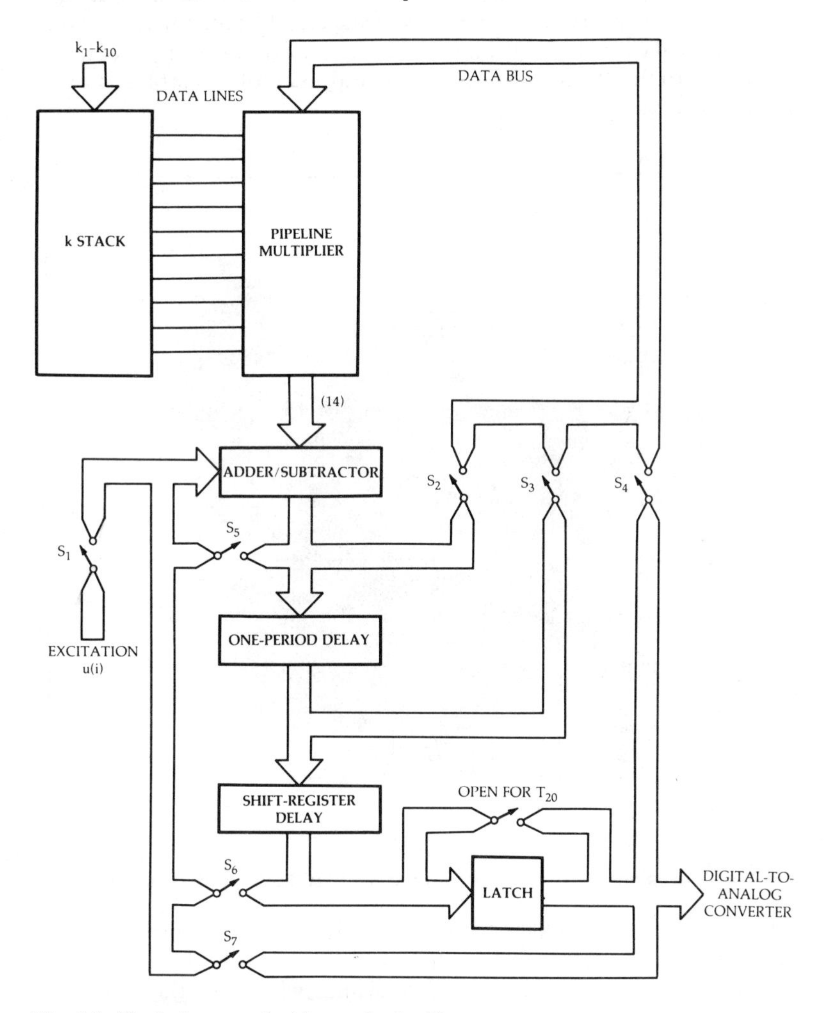

Fig. 5-5. Block diagram of a 10-stage lattice filter.

THE TMC 0270 CONTROLLER

The third chip of this three-chip speech synthesis system is the TMC 0270 controller. It is a slightly modified calculator chip that is related to the TMS 1000 microprocessor family.

During speech, the TMC 0280 synthesizer accesses the phrase ROM directly, but when it receives an end-of-phrase command, it returns control to the controller. In periods of silence, the controller has complete control over the ROM interface lines; it can transfer phrase start addresses to the ROM or it can access address look-up tables or other auxiliary data from the ROM.

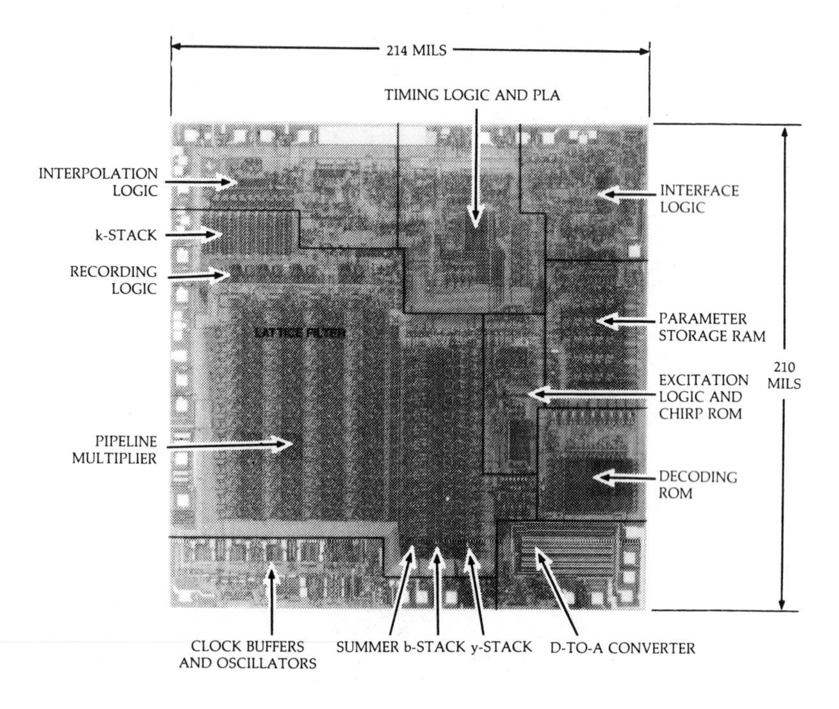

Fig. 5-6. Microphotograph of Texas Instruments' TMC 0280 speech synthesizer.

There are five special lines from the controller to the synthesizer that transfer data and commands within the system. One of these lines is the processor data clock. It is used to determine when the other four lines are valid.

The controller has several commands. In addition to the normal Read command, there is the Speak command. This tells the synthesizer to start accessing data and begin speaking. A Test/Talk command allows the controller to determine if the synthesizer has finished talking. The controller can tell the ROM to read 1 bit and the ROM will shift a 1 bit into a 4-bit shift register. This sequence can be repeated three more times until 4 bits are ready for output into the four control and data lines.

The TMC 0280 speech synthesizer was designed so that most microprocessors can be used as controllers. Fig. 5-6 is a microphotograph of the TMC 0280 speech synthesizer. It is now commercially designated as the TMS 5100.

CHAPTER 6

SYNTHESIS-BY-RULE AND TEXT-TO-SPEECH SYSTEMS

The ability to convert printed text into a spoken message is one the primary objectives of electronic speech synthesis. At the present time, the best results have been obtained by using a stored vocabulary of data defining the parameters of the specified elements of human speech. This vocabulary may represent phrases, words, syllables, or fractions of syllables. A computer is used to select elements from the stored vocabulary and join them together smoothly and naturally.

It has always been easier to join words and phrases than syllable-length elements and even syllable-length elements can be joined together more satisfactorily and easier than phoneme- or allophone-length elements. Less computer calculation is needed to join word and phrase elements because the normal vocal coarticulation remains intact within words and phrases. At the phoneme and allophone level, there are many more coarticulation options. This has led to one of the important tradeoffs in electronic speech synthesis.

SPEECH SYNTHESIS

The variety of messages that can be synthesized for a given amount of memory is greater for allophones and phonemes than it is for words or phrases but the quality of the message formed from stored words or phrases is generally superior.

Research is being carried out at Bell Laboratories on the pure computational synthesis of speech based on the fundamental principles of speech. No stored human-speech elements are being used as in the commercial products presently available from Texas Instruments, Votrax, and others. The complexity of Bell Laboratories' processing approach is far greater than is likely to be used in any commercial products or procedures in the foreseeable future. Nevertheless, in its present state of development, the quality of the voice from the Bell Laboratories' apparatus is robot-like. The computational power of at least a minicomputer is necessary for processing. However, text-reading machines for the blind are being designed on a basis of computer synthesis.

All of the text-to-speech systems developed to date have involved tradeoffs in versatility, quality, complexity, and cost. As the cost of digital memory continues to drop, speech synthesis systems that store thousands of parametric vocabulary elements, both word size and syllable size, are becoming increasingly economical.

STORED SPEECH

There is, however, a fundamental difference between the approach to speech synthesis of the major semiconductor manufacturers like Texas Instruments and National Semiconductor and that of the Bell Laboratories, although all these organizations appear to be working along the same lines. The objective of the semiconductor companies is to produce semiconductor devices—synthesizer chips, memory chips, microprocessors, and other ancillary devices which can be sold to consumer, commercial, industrial, and, perhaps, even military system OEMs. By contrast, the Bell Laboratories programs are aimed at reducing operating costs and making their telecommunications systems more effective and efficient.

Computer-like processors with some speech capability already link electromechanical and electronic switching systems with telephone customers. The *Voice Storage System's* processor interacts with switching systems to receive, store, and send customer messages automatically. The *Automatic Intercept System* handles calls placed to nonworking numbers; it issues recorded announcements that tell why a number is not in service and provides the new number if it is available. Experimental automated directory-assistance computers look up the telephone numbers and speak them to customers. *Automated Coin Toll Service* handles many toll calls that are made from coin stations. It announces such information as the amount of money that must be deposited. However, in another application of as much value to industry as to telecommunications, computer-stored spoken instructions now help technicians wire and test telephone equipment. The concept could be applied to virtually any complex assembly-and-test task that calls for the craftpersons to make constant references to instructions and diagrams.

In most of these telecommunications applications, the spoken messages are selected from a limited number of words and phrases spoken by a person. These are recorded, digitized, and stored as "waveforms" in the computer. Output messages are formed by accessing the waveforms for the stored words and phrases and, then, assembling or coarticulating them to form a meaningful message.

This approach is effective for messages that have no complex contextual relationships. A good example is spoken digits, such as phone numbers. The system's fluency and versatility or range of messages is limited. Except for adjustments of the silent intervals between words and phrases, they are used exactly as recorded. It is not possible in most of these systems to smooth transitions between segments or to vary the voice inflection to suit the context. Moreover, the number of vocabulary elements from which the computer-based system can piece together a message is limited by its digital storage capacity.

Nevertheless, the objective of both commercial electronic speech synthesis and the telecommunications industry research and development is the perfection of techniques to produce speech of a quality that is acceptable to listeners while using a minimum of both processor power and digital memory. While the option of virtually unlimited messages is open to major telephone utilities with on-line computers, the industrial applications will tend to be limited to a restricted vocabulary that is more appropriate to the end use.

SPEECH GENERATION

The main stages in creating a synthetic message are similar to those of human speech (Fig. 6-1). The printed text is first converted, according to the conventions of the language employed, into a sequence of sym-

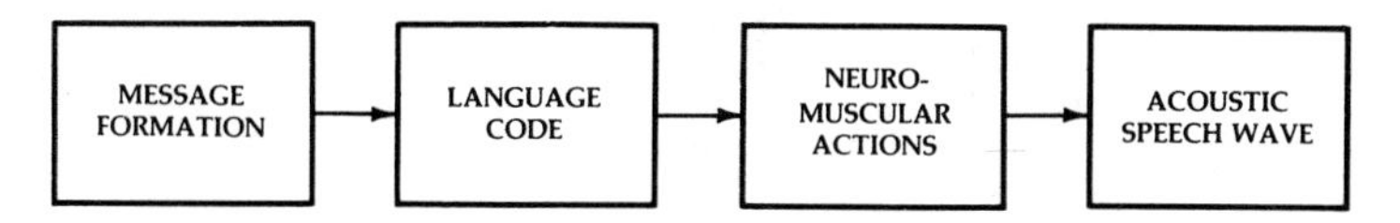

Fig. 6-1. Formation of human speech.

bols representing the distinctive speech sounds—phonemes or allophones (Fig. 6-2). This can be done by algorithm for letter-to-sound conversion. An alternate method is by look-up of each text word in a stored pronouncing dictionary. More typically, a mixture of these techniques is used, with the size of the stored dictionary and the sophistication of the letter-to-sound rules present in an inverse ratio.

One currently existing computer-based approach at the Bell Laboratories uses a 1000-rule algorithm and a 1000-word dictionary (to accomplish pronunciations that cannot be handled by a rule). In contrast, another system uses a 40-rule algorithm and a 30,000 word dictionary. These systems give comparable performance for the synthesis of contextual material.

There are two distinctly different methods for transforming a discrete phoneme or an allophone sequence into the series of continuous parameters that are needed to control a speech synthesizer. The first and most commonly used method is based on a stored vocabulary of human-derived parameters. These have been measured from sentences, words, or syllables that have been stored in some form of memory. This is illustrated by Method 1 in Fig. 6-3 where the human-derived parameters are located in Box B.

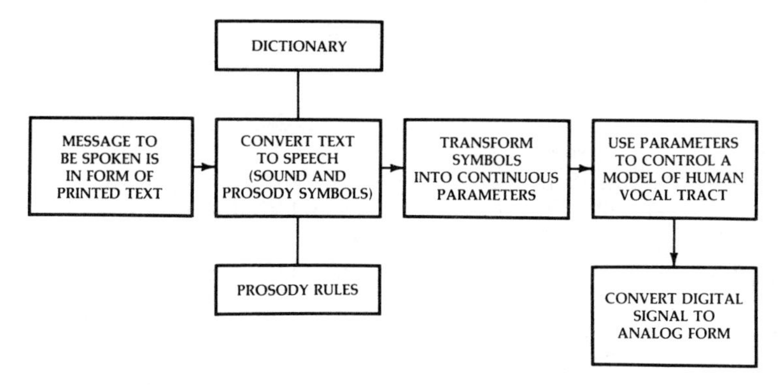

Fig. 6-2. Computer-based text-to-speech method of creating synthetic speech.

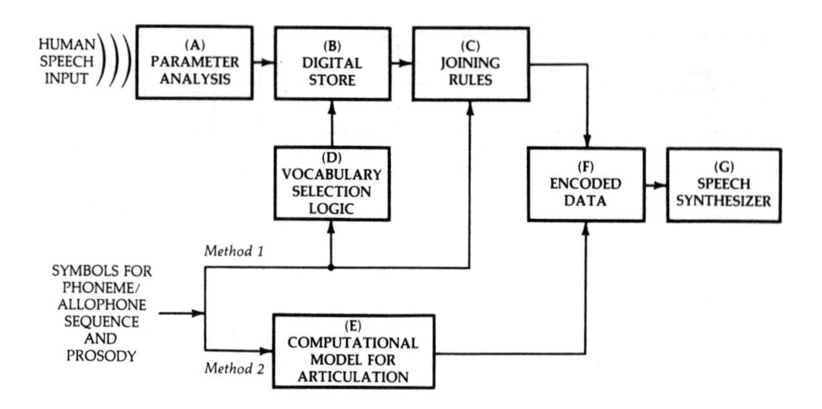

Fig. 6-3. Synthetic speech can be derived from the human voice (Method 1) or it can be computed mathematically.

The lengths of the vocabulary elements correspond either to words, syllables, or fractions of syllables (phonemes or allophones). The voice-synthesis system requires a digital storage capacity of from about 500 to 2000 bits per second of speech in order to store these parameters. This is far less than the digital rate required to store the original waveform. (As stated in Chapter 1, it is typically 24,000 to 64,000 bits per second.)

The phoneme or allophone sequence generated by the sequence of the input message determines which vocabulary elements (Box D) will be accessed from memory and their order. Once the elements are accessed, the corresponding parameters must be smoothly joined. This is done by the computer. The input symbols specify whether the speech sounds are "voiced," "unvoiced," or a combination of both.

The computer also determines the accompanying prosody—the variations in voice pitch, intensity, and sound duration that are appropriate to the context. Prosody determines, for example, whether the speech is declarative, interrogative, or exclamatory. Based on the prosody, the computer generates, by rule, the additional continuous parameters needed to control voice pitch and sound intensity. It must also modify the duration of the selected vocabulary elements.

In cases where the stored vocabulary is composed of word-length elements, the conversion from text-to-speech sound is simple and direct. However, it may be necessary to adjust for prosody.

In the most sophisticated method for transforming the discrete phoneme sequence into continuous speech parameters, no links to the originally recorded human speech are used (Method 2). The human speech parameters are calculated completely by programmed rules. The rules describe the fundamental principles of speech sound generation and they, typically, include a dynamic model of articulation. The calculated parameters are then issued to the electronic synthesizer, which responds to their control in the same manner as it responds to parameters derived from the stored vocabulary method. The purely computational approach is more complicated and requires more processing power in the computer than the other methods.

The basis of the calculations is a mathematical description of the active features of the human vocal tract. They include a programmed description of the physics of sound as they relate to the motions of the mouth, jaw, tongue, velum, and larynx, as described in Chapter 3. At the present time, Bell Laboratories is actively engaged in this research and their system is still not perfected.

The digital data derived from the calculated method, as well as the human-derived vocabulary method, is used to calculate the prosody parameters. It should be noted, however, that the complete set of parameters (the encoding of both the sound spectrum and the prosody) is similar for both techniques.

CHAPTER 7

TEXAS INSTRUMENTS PRODUCTS AND TECHNOLOGY

Texas Instruments Incorporated has been continually adding to its family of speech synthesis integrated-circuit products—both synthesizers and read-only memories (ROMs). In this chapter, the devices that are now available will be described briefly.

VOICE SYNTHESIS PROCESSORS

Texas Instruments refers to its synthesizer ICs as voice synthesis processors or VSPs. At the present time, it is offering three VSPs: the TMS 5100, the TMS 5110A, and the TMS 5220. Speech encoding on all TI VS processors is achieved with LPC coding. The codes for twelve synthesis parameters (pitch, energy, and ten filter coefficients) serve as inputs to the VSP. After decoding by the VSP, the output codes become time-varying descriptions of the LPC model of the original voice.

Two kinds of inputs are applied to the digital filters of the VSPs: periodic and random. Voiced sounds that have a definite pitch, such as vowels or voiced fricatives (z, b, and d), are reproduced from periodic inputs. By contrast, unvoiced sounds, such as s, f, t, and sh, are derived from random inputs. The voiced and unvoiced excitations are generated by separate sources. The output to drive the speaker is obtained from the filtered input to the digital-to-analog converter.

Texas Instruments' VSPs operate at an average data rate of 1200 bits per second, which is low when compared to the 100,000 bits per second necessary for direct speech digitizing. A TMS 5000-series VSP, when teamed with a TMS 6100 128K ROM, can "speak" about 200 words in more than 100 seconds. A single TMS 5100 or TMS 5220 can interact with as many as sixteen 128K memories. This combination can provide about 3200 words of speech. Moreover, the TMS 5220 is also able to tap host storage or off-line storage to produce a virtually unlimited number of words.

TMS 5100 and TMS 5110 VSPs

The TMS 5100 voice synthesis processor is the commercial version of the PMOS VSP TMC 0280 speech synthesizer described in Chapter 5 and which is used as the heart of the *Speak and Spell* ™ product. It interfaces easily to 4-bit microprocessors such as the TMS 1000 family of PMOS/CMOS microcomputers. A complete voice synthesis system can be assembled from three devices, a speaker, and a keyboard. This is shown in Fig. 7-1. These devices are the TMS 5100 voice synthesis processor, the TMS 6100 voice synthesis memory, and the TMS 1000 microcomputer.

Speech is synthesized when the TMS 5100 processes a variable data rate stream of encoded speech data and converts the results into an audio signal with its on-chip 8-bit digital-to-analog converter and push-pull amplifier. The TMS 5100 was designed to work with up to sixteen TMS 6100 128K voice synthesis memories. All the necessary control signals are produced by the VSP. In typical circuits, control of the TMS 5100 VSP is external through four pins and a command or CPU clock. The VSP runs at a low average data rate of 1200 bits per second. In addition to being able to process male, female, and children's voices, the TMS 5100 can generate tones, chimes and sound effects.

The TMS 5110 has characteristics that are similar to the TMS 5100. It also increases voice fidelity over the TMS 5100 and it has an optimized coding table for general voice synthesis.

TMS 5220 VSP

The TMS 5220 voice synthesis processor permits the generation of synthesized speech under the control of 8- and 16-bit microprocessors. It is designed to interface with most microprocessors over an 8-bit bus.

Most of the input/output functions and processing operations necessary to produce speech are performed by the processor itself. The VSP behaves as an attached processor to the host CPU and performs its synthesis tasks when appropriate commands are sent by the CPU. Speech synthesis can be added to existing microprocessor-based systems because the demands made on the host microprocessor are slight.

Speech data can be stored in up to sixteen TMS 6100 voice synthesis memories, in host-system storage, or off-line in a memory disk. An on-chip first-in/first-out (FIFO) 16-byte buffer "buffers" two frames of speech data from the host microprocessor. A simplified block diagram of a basic system that is built around the pin-compatible TMS 5200 VSP is shown in Fig. 7-2.

An advanced coding table is responsible for the improvement in voice quality and frequency response, as well as the musical tones and sound effects, that can be obtained from the TMS 5220. Again, as in the TMS 5100 VSP, the average rate is 1200 bits per second. The TMS 5220 VSP can string allophones to form words, phrases, and sounds, and the ability to produce women's and children's voices is said to be improved.

VOICE SYNTHESIS MEMORIES

In addition to the VSPs, Texas Instruments offers a series of VSMs or voice synthesis memories. Two of these are discussed here.

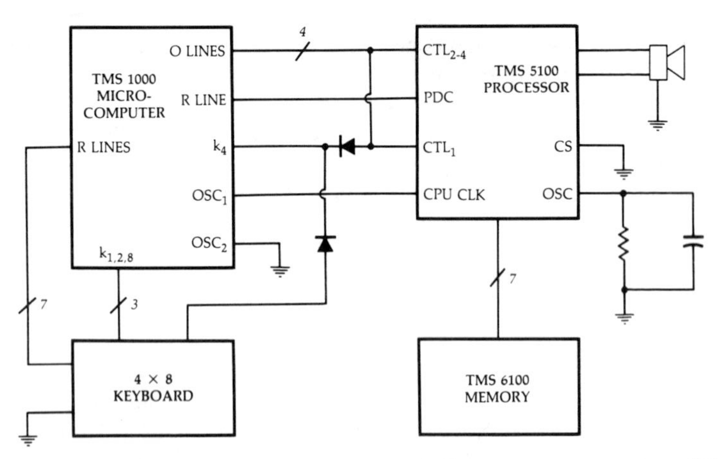

Fig. 7-1. Block diagram of a TMS 5100 voice synthesis processor system using a TMS 1000 microprocessor.

TMS 6100 VSM

The TMS 6100 voice synthesis memory is a PMOS 128K read-only memory that is internally organized as 16K-by-8 bits so as to adapt it for use with the TMS 5100 series and TMS 5200 series of VSPs. It uses a multiplex addressing scheme and has an internal 14-bit address counter/register. No address decoding is required for up to 16 memories. The on-chip memory-address register autoincrements through speech data until it is stopped by the VSP. Look-up and branch capability permits random access to the speech data on the chip. Fig. 7-3 is a simplified block diagram of the TMS 6100 VSM.

TMS 6125 VSM

The TMS 6125 voice synthesis memory is a PMOS 32K-bit ROM, internally organized as 4K-by-8 bits. The voice synthesis memory employs a multiplex addressing scheme with an internal 14-bit address counter/register. Twelve bits of the address go directly to the ROM array. The remaining two bits address two programmable gates to select one of a possible four voice synthesis memories. Because of the on-chip address decode logic, no address decoding is required for up to four voice synthesis memories. The on-chip memory address register autoincrements through the speech data until it is stopped by the voice synthesis processor. Look-up and branch capability is another feature that allows vocabulary-independent access to speech data on the chip.

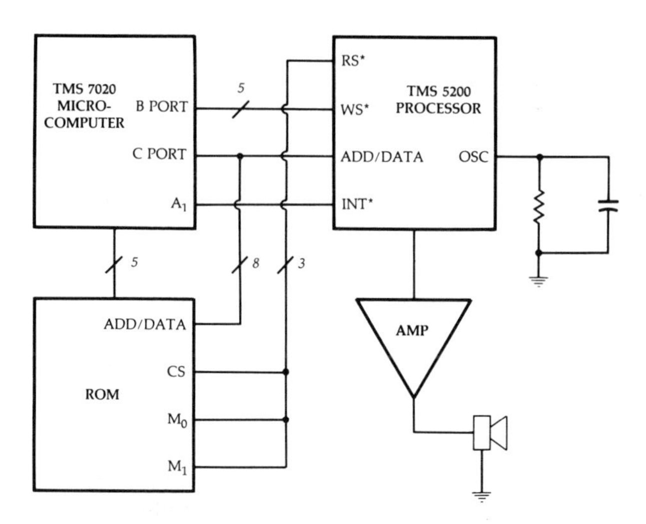

Fig. 7-2. Basic VSP system based on the TMS 5200 voice synthesis processor.

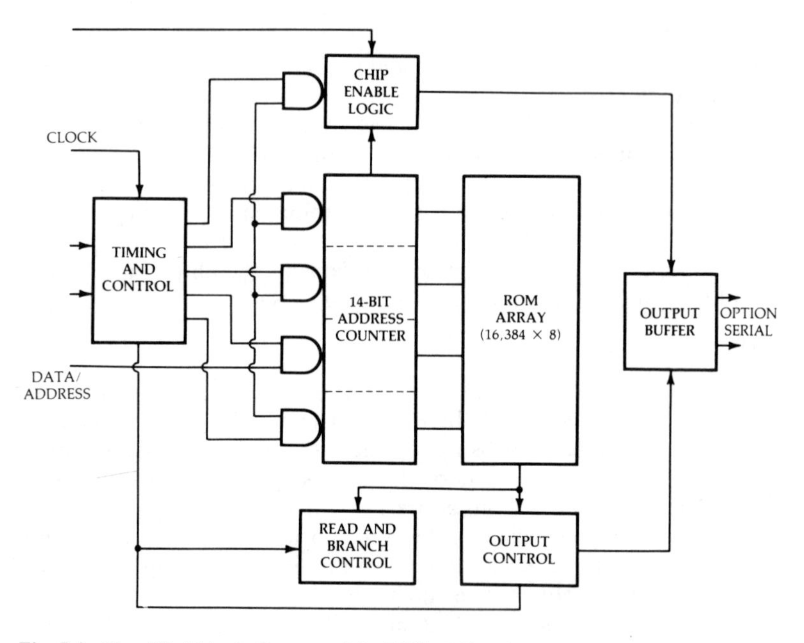

Fig. 7-3. Simplified block diagram of the TMS 6100 voice synthesis memory.

VOICE SYNTHESIS EVALUATION KITS

Speech evaluation kits are designed to permit users and designers to evaluate Texas Instruments' voice synthesis capability on 4-, 8-, and 16-bit microprocessor systems. Several kits are discussed in the following paragraphs. Each kit contains all the necessary hardware and a complete user's manual that covers interfacing and software design.

TMSK202 Evaluation Kit

The TMSK202 is a TMS 5220 speech evaluation kit that is 8-bit and 16-bit microprocessor compatible. The TMSK202 kit contains a TMS 5220 voice synthesis processor, three preprogrammed memories, and all the necessary data sheets and manuals needed to provide background technical information. The first of the three memories is the VM81002 32K-bit EPROM which is preprogrammed with 35 words and phrases for male voice rendition. The second, a VM61002 ROM, stores 206 words of industrial vocabulary for male voice rendition, while the third, a VM71003 ROM, stores data for 34 words that are rendered in a female voice. Each of the words, syllables, and letters stored in memory may be accessed individually to permit concatenation in the formation of phrases and sentences. A simplified block diagram of the TMSK202 system is shown in Fig. 7-4.

TMSK101 Evaluation Kit

The TMSK101 speech evaluation kit, like the TMSK202 speech evaluation kit, is also intended for designers who wish to evaluate TI's voice synthesis products except that, in this case, it is limited to control by 4-bit microprocessors. This evaluation kit contains a TMS 5100 voice synthesis processor and a VM61001 128K masked ROM that has an industrial vocabulary. The VM61001 contains 204 individually encoded words and alphabetical letters that may be accessed individually and, then, concatenated to form phrases and sentences. The VM61001 contains sequencing logic as well as an indirect addressing capability to minimize loading on the host microprocessor. A simplified block diagram of the TMSK101 architecture is shown in Fig. 7-5A.

TMSK101A Evaluation Kit

The TMSK101A speech evaluation kit, like the TMSK101, is intended for the evaluation of the TMS5100 voice synthesis processor. It is also 4-bit microprocessor compatible. The kit contains a TMS 5100 voice synthesis processor, three memories that are preprogrammed with speech data, and the complete user's manual which provides the neces-

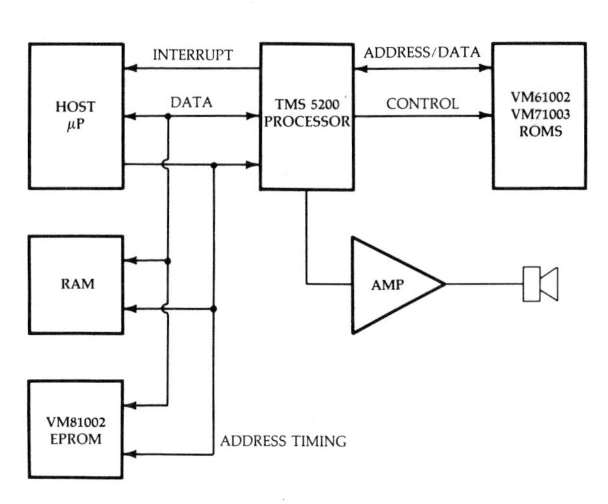

Fig. 7-4. Simplified block diagram of Texas Instruments' TMSK202 voice synthesis evaluation kit.

sary background information on interfacing and software design. A VM61001 ROM, the first of three memory devices, stores data for 204 individually encoded words and letters that may be accessed individually. These industrial vocabulary words are then concatenated to form phrases and sentences and are rendered in a male voice. The second memory, a VM71001 ROM, stores data for 50 speech elements that are to be spoken in a male voice, while the third memory, a VM71002 ROM, stores data for 35 of those elements for a vocabulary that is spoken by a male voice. A simplified block diagram of the TMSK101A voice synthesis evaluation kit is shown in Fig. 7-5B.

TMS5100 Series Evaluation Board

The TMS5100 speech evaluation board is a hand-held, battery operated, demonstration unit that contains a TMS 5100 voice synthesis processor, a TMS 2532 32K EPROM, a PROM that contains instructions for the processor, and an 8-ohm speaker. The TMS 2532 EPROM contains eight phrases of synthesized speech, each about three seconds long. More EPROMs can be added to increase the number of phrases. Customers can choose words and phrases from a wide selection.

The TMS 5100 evaluation board is intended for use in the evaluation or demonstration of Texas Instruments' voice synthesis devices. The user

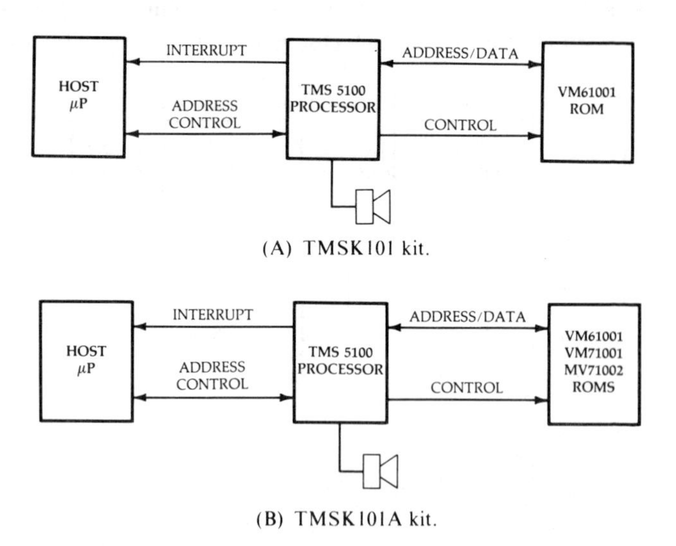

(A) TMSK101 kit.

(B) TMSK101A kit.

Fig. 7-5. Block diagrams of the system architectures for two voice synthesis evaluation kits from Texas Instruments Incorporated.

selects the desired phrase with the use of a rotary dial. The board can also be interfaced to a microprocessor by using a buffer circuit. Thus, it can provide a limited number of "canned" voice synthesis phrases for those products that can be improved with the addition of voice synthesis. Fig. 7-6 shows a simplified block diagram of the TMS5100 Series Evaluation Board. Fig. 7-7 is a photograph of the actual board.

MEMORY PROGRAMMING

Texas Instruments uses two approaches to programming memories for use in speech synthesis. Both approaches share the same process equipment although one does require additional preparatory equipment. The two approaches are intended to meet various speech application requirements and they offer the user an opportunity to relate speech quality to the cost of software preparation and the cost of hardware in the OEM product—particularly memory requirements.

The speech composition and editing system starts with text entered from a keyboard. This system uses an integrated set of interactive processes for editing text, phonemes derived from that text, allophones derived from the phonemes, and LPC conversion. The output of this process is data formatted on an erasable field-programmable memory, or EPROM.

A large investment in computer equipment is needed to carry out the process. Included in the system is a minicomputer with disk memory, a PROM programmer, a plug-in board for speech audition, and a quality crt editing terminal. In addition, there is a software package associated with the operation of the minicomputer system and separate hardware and software that permits text editing on the crt terminal.

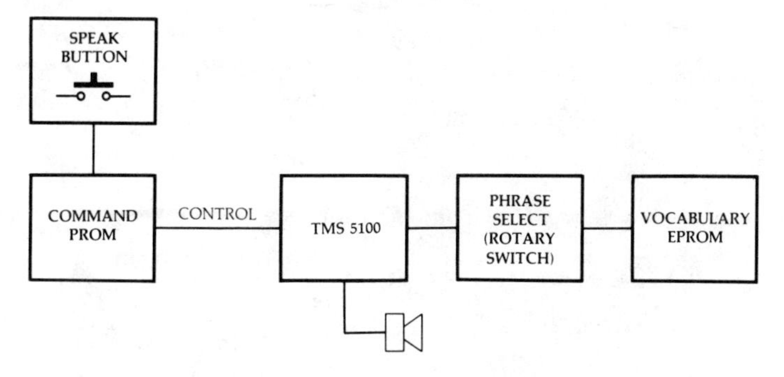

Fig. 7-6. Block diagram of the TMS510 Series Evaluation Board by Texas Instruments Incorporated.

The collection/LPC analysis system calls for sources in the form of recordings of spoken words, phrases, and sentences. Intensive LPC analysis is then performed on this recorded material before it is processed with the computer-based equipment listed earlier. As in the case of text-to-speech data, the LPC-analyzed data is edited, played back for review and alteration, and, finally, used to program an EPROM. The data in the EPROM can then be re-recorded and re-edited, if necessary, following listening tests.

The collection/LPC analysis processing also calls for an audio system to collect and record speech data, as well as software to perform LPC analysis.

The speech-composing/editing (or text-to-speech) system starts with speech that is entered on the keyboard and allows the editing of typed-in text on four levels—text, phonemes, allophones, and LPC binary code. First, the text can be edited as it is entered. It is then converted to a phonemic-alphabet representation according to a comprehensive set of rules. The phoneme output that is generated will have an accuracy of 90% to 92%.

The next conversion changes the phonemes to allophone notations which are more detailed variants of phonemes. These include data based on word context such as pitch, timing, relative word placement, intonation, and other linguistic considerations. The allophone nota-

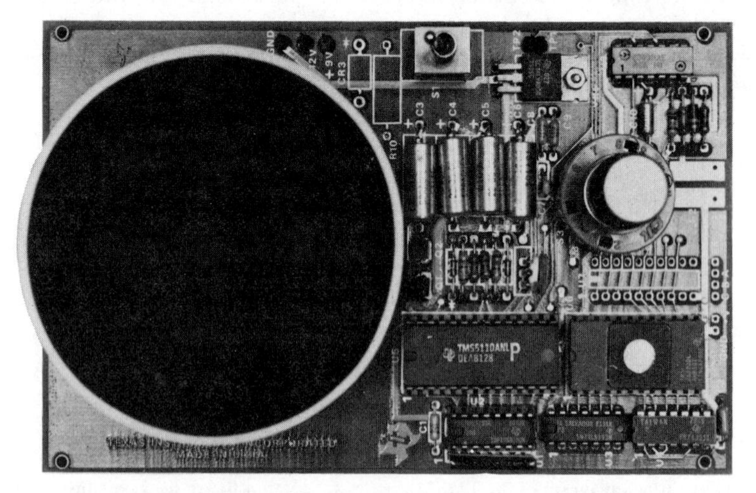

Fig. 7-7. Texas Instruments' TMS5100 Series Evaluation Board.

tions are then converted to Texas Instruments' LPC-10 code which, in binary form, can drive a speech synthesizer chip like the TMS 5220 and produce an audible speech output. Even after this output is obtained, it may be further refined by successive listening and editing until the desired natural-sounding synthetic speech is achieved.

The LPC display is the most complex of the four. A code in LPC-10 has 12 parameters in a frame. These include energy, pitch, and ten reflection coefficients (K_1 to K_{10}) which are represented by 50 bits.

The editor can interact at any of the four levels. Those editors without a linguistic background will probably want to work only at the text level. This will typically be done by creative misspelling of words. Those with linguistic training may prefer to work at the phoneme or allophone level.

After the editing process has been completed and the speech is found acceptable, the file data are formatted for EPROMs for use in the end product.

An average word is four or five allophones long. To reduce the memory requirements, one PROM can be programmed to carry all the allophones (approximately 128) and a second PROM can be used to select the sequence in which they are used. This permits the assembling of a very large vocabulary with minimum storage.

If only a few words are to be synthesized, all the needed allophones can be serially arranged in one small PROM and, then, read out sequentially exactly as they are concatenated in the development system.

The collection and analysis of acoustic-sourced speech calls for an audio system for collecting and recording speech as well as for software for the LPC analysis. As indicated earlier, it still depends on the text-to-speech equipment for subsequent editing and PROM formatting. While the text-to-speech minicomputer does not need to be a powerful machine, it must have a substantial memory. However, the minicomputer used for LPC analysis must be capable of high-performance.

MULTI-AMPL™ SPEECH DEVELOPMENT SYSTEM

The speech development system from Texas Instruments permits

Multi-AMPL is a trademark of Texas Instruments Incorporated.

equipment manufacturers with the in-house equipment to develop, analyze, and edit those synthesized speech vocabularies that they wish to customize for their products. Users can control each step of the vocabulary development from initial script preparation to the programming of speech data on an EPROM.

The voice synthesis development system uses two different methods to generate speech data. In the analysis synthesis method, data from natural speech is used to estimate the parameters of a time-varying linear model. Then, the synthesis process creates the synthetic speech signal from these parameters. In the constructive synthetical method, the words, phrases, and sounds are built from a library of basic speech sounds.

Two principal hardware components form the Multi-AMPL speech development system. The TMAM9080 computer is a high-performance general-purpose computer for speech data processing. It controls the activities of the data collection processor, which is the second important unit of hardware. These activities include record, playback, save, and load. In addition, the system provides the linear predictive coding (LPC) synthesis output channel, EPROM programming facility, an on-line storage device, and a tape drive for off-line storage of speech data.

The data collection processor (DCP) is an intelligent "slave" device to the TMAM9080 and provides the analog interface required for the collection and playback of speech signals. All circuitry is solid state. The DCP is designed to be positioned near a microphone or other source of speech input.

In addition to the major items of equipment described, a Multi-AMPL speech development system calls for microphones, mixers, audio amplifiers, loudspeakers, and audio interconnect cable. A listing of Multi-AMPL equipment is given in Appendix B.

System Hardware

Multi-AMPL development systems can be configured with a wide range of processor, disk, tape, and memory options. The TMAM9080 system for production operation includes a 990/12 CPU, a 50-MB disk, a 1600-bpi tape drive, a 256K memory, a 3K cache memory, and two Model 911 video-display terminals. Fig. 7-8 shows a basic speech development system diagram.

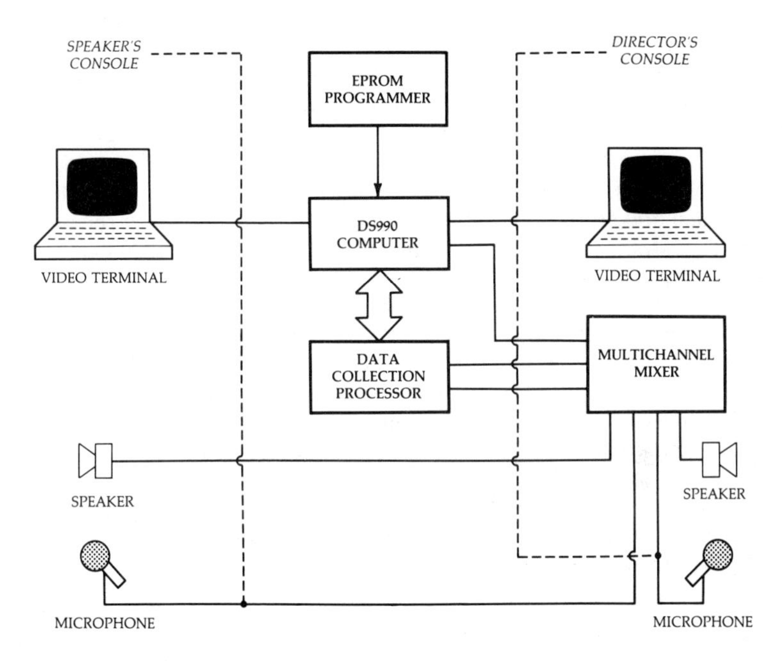

Fig. 7-8. A basic speech development system.

The minimum equipment complement includes a voice synthesis processor audition board, a PROM programmer kit, a PROM personality card, DXAMPL PROM control software, and the speech-composition system software package. For recording and analysis of speech signals, the system requires the data collection processor, a high-speech parallel interface card, and the speech-analysis system software package.

Optional equipment that adds to the efficiency and flexibility of the system includes a hard-copy printer, an expansion memory, and expansion terminals.

System Software

The software components of the Multi-AMPL speech development system consist of the script creator, the speech composition system editor, the recording session control and analysis package, and the speech linker. These software programs run within the Texas Instruments DS990 DX-10 operating system to provide multiuser, multitasking support for the speech development system software.

The software packages are designed to provide interactive interface responses that will lead users through vocabulary generation quickly and easily without intimidating them. The first step in developing synthesized speech is the preparation of the script of desired words, phrases and sound effects. The next step is to make the choice between constructive synthesis and analysis synthesis for producing the speech data. The resulting speech derived from the selected method are then auditioned and edited. Finally, the vocabulary development process is completed when the edited data are converted to a format that can be accepted by an EPROM.

A data base management system within the speech development system preserves the character of files, interacts conveniently with all utilities, and provides a method for filing the processes and information within the data bases. This data-base resource is important because speech processing involves a large amount of data generation.

The final step in the use of the development system is the programming of the EPROM from the software. The user can write speech data into EPROMs in different formats and can then choose the one that is easiest to adapt to the end-usage system.

PREPROGRAMMED SPEECH

Texas Instruments has developed a line of preprogrammed ROMs that include selected application-oriented vocabularies. These vocabularies are used in industry, avionics, military operations, and, also, for stating numerical values, time, and the weather. Both male and female voice intonations are available in some of the vocabularies. The vocabularies are intended for use with Texas Instruments' TMS 5100 and TMS 5220 voice synthesis processors.

Vocabulary ROMs

Table 7-1 lists those vocabulary ROMs that were available from Texas Instruments during 1982. They are said to be useful in applications where only small quantities of a product will be made or when used for evaluation and prototyping systems.

Three vocabular ROMs are offered for industrial applications. The VM61001 and VM61002 ROMs each have vocabularies of approximately 200 words spoken in a male voice. The VM61001 ROM is compatible with the TMS 5100 VSP while the VM61002 ROM is

Table 7-1. Preprogrammed Speech Vocabulary ROMs

VOCABULARY	TMS5200 COMPATIBLE	TMS5100 COMPATIBLE
Industrial	VM61002 (206 words—male)	VM71001 (50 words—male) VM61001 (204 words—male)
Numeric/Time	VM71003 (35 words—female)	VM71002 (35 words—male)
Avionic	VM61005 (134 words—male)	
Weather/Clock	VM61003 (138 words—male)	
Military	VM61004 (147 words—male)	

compatible with the TMS 5220 VSP. A smaller industrial vocabular ROM of about 50 words, also spoken in a male voice and designated the VM71001, is compatible with the TMS 5100 VSP.

When numbers have to be spoken, the VM71002 and VM71003 ROMs offer about 35 words and phrases that are appropriate. While both offer the same vocabularies, the words of the VM71002 ROM are spoken with a male voice; it is compatible with the TMS 5100 VSP. However, the words of the VM71003 ROM are spoken in a female voice and that memory is compatible with the TMS 5220 processor.

Where words appropriate to meteorology and the telling of time are desired, the VM61003 ROM is available; it contains 138 words that are related to those subjects and spoken in a male voice. The VM61004 ROM contains a vocabulary of 147 military words which are spoken in a male voice. A vocabulary for aircraft communications is contained in the VM61005 ROM. It consists of 134 words spoken in a male voice. The VM61003, VM61004, and VM61005 memories are all compatible with the TMS 5220 voice synthesis processor.

Both the TMSK101 and TMSK101A voice synthesis evaluation kits include VM61001 vocabulary ROMs. The TMSK101A voice synthesis evaluation kit contains a VM71001 and VM71002 ROM as well. Appendix C contains word, sound, and letter lists for the VM71001, VM71002, VM71003, and VM61002 ROMs.

Speech Library Service

Texas Instruments offers pre-encoded words and phrases from its speech data base at lower cost per word than custom-processed data. Individual words or phrases obtained from the TI Speech Library

Service are equivalent to custom-processed words or phrases. However, there is a danger that the stringing of library words and phrases together without any regard to their original context could result in speech that lacks the quality of custom-processed phrases. This could occur because the library words were originally recorded, processed, and edited for various specific applications.

Additional editing of library words can smooth out the stringing together of phrases to some extent, but where natural speech and inflection are important, custom processing still remains the best approach. However, the use of library speech is an attractive option for all prototypes and for many end-use applications.

CHAPTER 8

NATIONAL SEMICONDUCTOR PRODUCTS AND TECHNOLOGY

National Semiconductor Corp. offers a speech synthesis system that consists of a speech processor chip (SPC) and a speech read-only memory. Together these components form a DIGITALKER™ kit. To this kit, a customer must add a clock input signal (or the necessary oscillator components), an audio filter and amplifier, and the control circuit function. Together, these units represent the minimum configuration. A block diagram of the configuration is shown in Fig. 8-1.

The maximum amount of directly addressable speech memory that is accessible by the speech processor chip (SPC) is 128K bits. However, external page addressing by the control circuit function can increase this ROM field as required.

THE SPEECH PROCESSOR CHIP

The SPC employs a time-domain synthesis technique. This technique reduces the amount of memory needed to store electronic speech by removing the excess or redundant data from the speech signal.

The four main techniques that perform the task are:

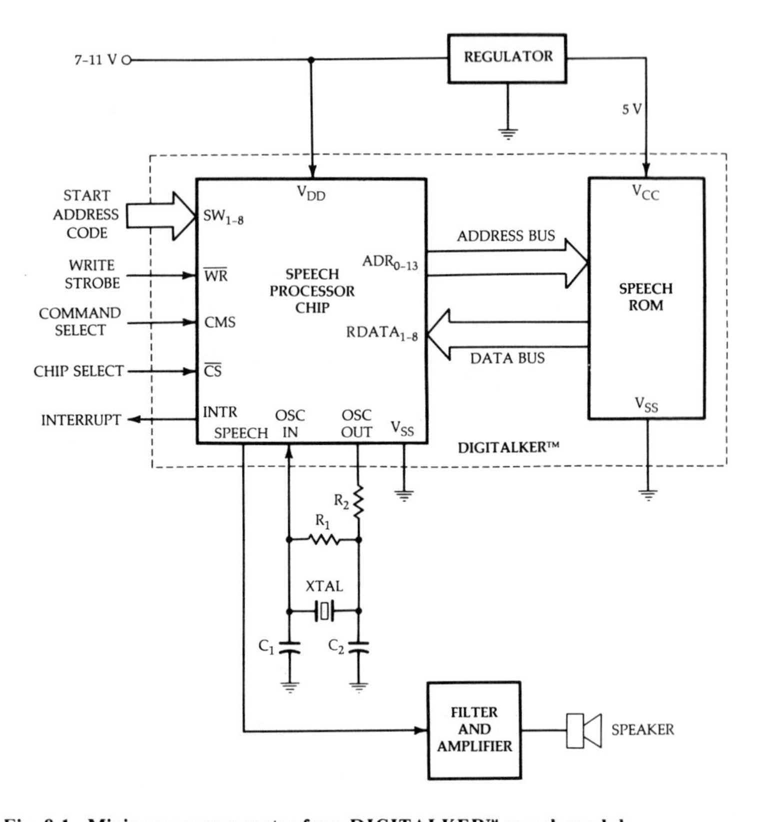

Fig. 8-1. Minimum componentry for a DIGITALKER™ speech module.

1. Elimination of redundant pitch periods.
2. Adaptive delta-modulation coding to minimize memory requirements.
3. Phase angle adjustments to create mirror image symmetry.
4. Replacement of the low-level portion of a pitch period with silence (half period zeroing).

To faithfully synthesize the original voice message, National Semiconductor Corp. uses an elaborate computer program to analyze a high-fidelity tape recording and, then, generate a ROM pattern that will accomplish the necessary action.

Fig. 8-2 is a block diagram of the SPC. The 8-bit start address bus allows up to 256 separately defined sounds or expressions to be stored in

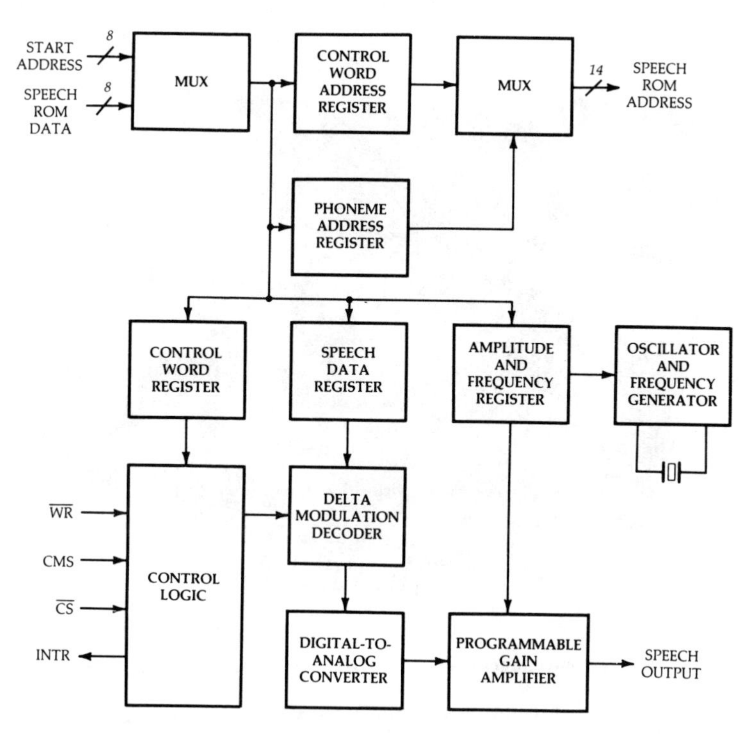

Fig. 8-2. Block diagram of a National Semiconductor speech processor chip.

the speech ROM. The control interface connected to the start address port can be either decoding logic, a MICROBUS™ port, or mechanical switches.

EVALUATION KITS

National Semiconductor offers some fully assembled voice synthesis evaluation boards for evaluating the operation and application of the DIGITALKER™ speech synthesis chip. One such board, the DT1000 evaluation board, shown in Fig. 8-3, requires only a single 9-volt power supply and an inexpensive speaker in order to provide a demonstration.

The DT1000 contains all the components required to produce speech on command: a speech processor chip (SPC), two speech ROMs (each containing 138 individual words), an output filter, an audio amplifier, a

DIGITALKER and MICROBUS are trademarks of National Semiconductor Corp.

Fig. 8-3. National Semiconductor's DT1000 DIGITALKER™ evaluation board.

keyboard, and a National Semiconductor COPS microcontroller with EPROM. The microcontroller contains stored data to provide the various functions of the board.

The two speech ROMs permit the user to link words consisting of numbers and the letters of the alphabet, assorted useful nouns, verbs, tones, and silence intervals into phrases and sentences. The DT1000 can also be interfaced easily to any external processor system with a standard 22-pin printed-wiring board edge connector. The DT1000 board measures 5 inches by 6 inches (12.7 cm x 15.24 cm).

Another evaluation kit, the DT1050, is available with the SPC and two 64K standard vocabular speech ROMs. These speech memories add 131 discrete words, spoken in a male voice, to National Semiconductor's speech synthesis lexicon.

DT1056/1057 kits expand National Semiconductor's DIGITALKER™ speech synthesis vocabulary to 274 standard words which are available in ROMs for a "hands-on" assembly of useful terms. New words that include "secure," "switch," "first," "floor," and "button" permit the

DIGITALKER™ to be used in many consumer and industrial applications. The vocabulary is stored in two 65K-bit ROMs.

The DT1056 is a complete kit that includes the DIGITALKER™ speech processor chip and speech ROMs. The DT1057 kit includes speech ROMs only. It is easily interfaced with National Semiconductor's DT1050 DIGITALKER™ kit, according to the company. Thus, the user has access to all 274 words.

The kits may be operated either with mechanical switches or with a microprocessor system. They require only simple amplifiers and speakers to produce speech.

CHAPTER 9

AMERICAN MICROSYSTEMS PRODUCTS AND TECHNOLOGY

American Mircrosystems, Incorporated (AMI) has introduced two speech synthesizers, an S3610 and an S3620, to meet differing end product requirements. The S3610, an LPC-10 speech synthesizer with an on-chip 20K speech data ROM, is configured in a 24-pin DIP package. The S3620 is basically the same device except that it does not contain an external speech data ROM. It is placed in a 22-pin DIP package. Both devices feature CMOS switched-capacitor filter technology, automatic power-down, single power-supply operation, direct loudspeaker drive, and 30 milliwatts of audio output power.

THE SPEECH SYNTHESIZERS

The S3610 speech synthesizer has a simple digital interface that consists of five word-select lines, a strobe input to load the address data and initiate operation, and a busy output signal. At the end of enunciation, the chip automatically goes into the power-down mode until a new word-select address is strobed in. The data rate from the speech ROM into the synthesizer is 2000 bits per second, maximum. Typically, the average data rate will be reduced to about 1200 bits per second with internal data-rate reduction techniques. This permits about 17 seconds of speech from the data on the internal ROM. The five word-select lines allow a maximum vocabulary of 32 words.

The digital interface circuitry of the S3620 synthesizer is compatible with most 4-bit and 8-bit microprocessors; it allows the processor to load data with or without a direct-memory access (DMA) controller. Loading occurs on a handshake basis. In the absence of a response from the processor, the synthesizer automatically shuts down and goes into a power-down mode. A busy signal allows the processor to sense the status of the synthesizer. The input data rate is also 2000 bits per second, maximum. Typically, however, the average data rate will be reduced to about 1400 bits per second with internal data-reduction techniques.

Synthesis in both devices is carried out with CMOS analog switched-capacitor filters operating at 8000 samples per second. Output interpolating filters and bridge power amplifiers give 30 milliwatts of audio output power from a 6-volt power supply. Both devices can be directly connected to 100-ohm loudspeakers and both have on-chip oscillators. External 640 kHz ceramic resonators are required for normal operation.

SYNTHESIZER OPERATION

The block diagram for an S3610 synthesizer is shown in Fig. 9-1. The diagram for an S3620 synthesizer is identical to the S3610 except that an 8-bit input latch replaces the word-decode ROM, the address counter, and the speech-data ROM. When the 8-bit latch shown in Fig. 9-1B is substituted for the components shown within the phantom area (dotted line box) of Fig. 9-1A, the chip is converted to an S3620 synthesizer.

The word-decode ROM of the S3610 speech synthesizer (in the phantom area of Fig. 9-1A) decodes the data presented on the word-select lines into the start addresses of the speech words stored in the speech-data ROM. Up to thirty-two 12-bit start addresses may be programmed into this ROM. The strobe signal selects the start address necessary to preset the address counter.

The address counter addresses the speech-data ROM. It is first preset to the desired start address and is then incremented each time a new byte of data is required for the synthesizer. The speech-data ROM contains

2560 bytes of LPC parameters encoded into a nonlinearly quantized packed format. This format permits the storage of each frame of LPC parameters in five bytes or less.

The end-of-word decoder detects the code that indicates the end of speech data and initiates the power-down routine after the previous frame has been enunciated. The rest of this description applies to both the S3610 and the S3620 synthesizers.

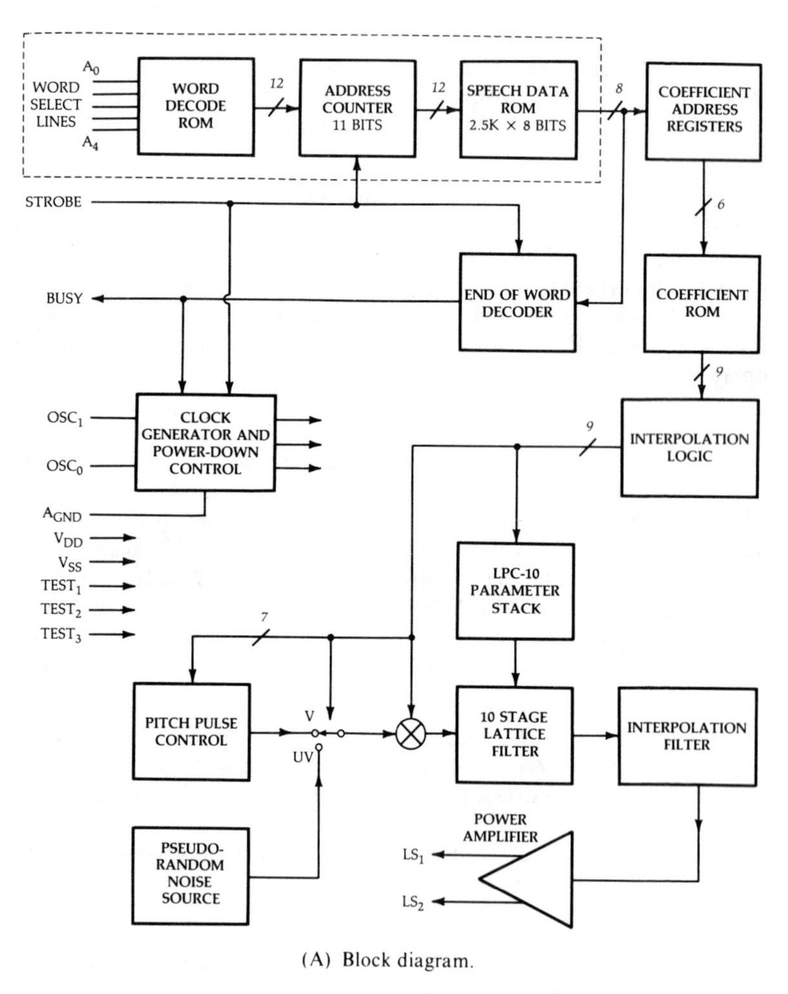

(A) Block diagram.

Fig. 9-1. The S3610 LPC-10 speech

Coefficient address registers assemble the data from the speech-data ROM into frames and, then, allocate the data into the thirteen parameters required for LPC-10 synthesis: pitch, gain, ten lattice filter coefficients, and the unvoiced/voiced speech division. The parameters are stored in an encoded format; decoding is done in the coefficient ROM. The coefficient address registers are used to store the assembled frame data and address this ROM. The coefficient ROM functions as a look-up table to decode the stored parameters into the LPC coefficients.

The coefficients for each frame of speech, normally 20 milliseconds, are interpolated four times during the frame by the interpolation logic in order to generate a smoother and more natural sounding speech. Thus, the interpolation period is one quarter of a frame period, normally 5 milliseconds. After interpolation, the coefficients are used to drive the pitch-pulse source, the lattice filter, and the gain control. Interpolation is inhibited during transitions from voiced to unvoiced speech.

The pitch signal source for voiced speech (vowel sounds) is generated by switched capacitor circuitry. Symmetrical bipolar pulses are generated at the rate specified by the pitch parameter.

By contrast, the pseudorandom noise source that represents unvoiced speech (fricatives and plosives) consists of a 15-bit linear code generator that has a periodicity of 32,767 sampling periods (4 seconds). The output of this generator is scaled to a lower value and is used as a random-sign constant-amplitude signal.

The voiced/unvoiced speech selector switch performs the function of alternately switching in the voiced or unvoiced signal sources as determined by the coding used to drive the filter during each frame. The

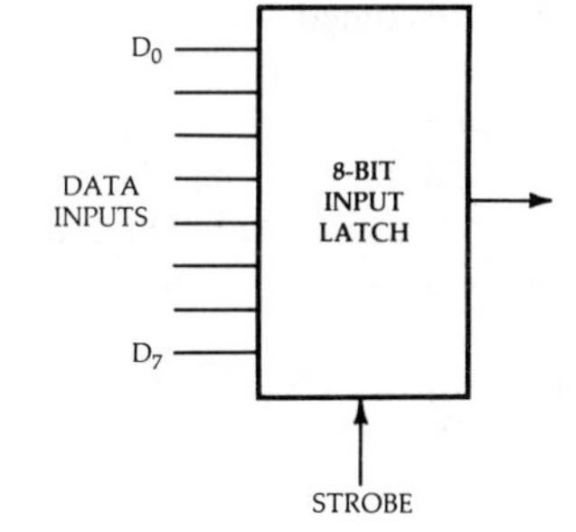

(B) 8-bit input latch.

synthesizer by American Microsystems, Inc.

LPC-10 parameter stack is a stack of ten filter coefficients that are used to control the lattice filter. The coefficients have an accuracy of 8 bits plus sign.

The 10-stage lattice filter simulates the effect on the vocal tract of the sound source (vocal chords). It is a switched capacitor 10-stage lattice filter that employs analog sampled data. The filter parameters are determined dynamically by the time-varying coefficients in the parameter stack. The filter operates at a sampling frequency of 8 kHz.

The output signal from the gain controller, which controls the signal level from the lattice filter to vary the sound level, is a sampled 8 kHz. An elliptic low-pass switched capacitor with a 160-kHz filter-sampling frequency filters out aliasing. It also conditions the output components and compensates for sin X-X distortion so the signal can directly feed the loudspeaker after amplification. The power amplifier raises the signal level to 30 milliwatts rms into a 100-ohm load. The output is a balanced bridge configuration.

Before interpolation, the input data rate of the synthesizer is 5400 bits per second (21,600 bits per second after interpolation). It consists of 12 parameters of 9 bits each, repeated every 20 milliseconds. A nonlinear quantization technique reduces the data rate to less than 2000 bits per second for storage.

Each of the 12 coefficients has a fixed set of values that is dependent on the speech data and is generated automatically in the analysis process. The parameters used to specify the coefficients are stored in the speech-data ROM and are used to address the coefficient look-up table ROM.

The speech data are further reduced by several techniques. For example, by reducing the order of the lattice filter from ten to four during periods of unvoiced speech, a 40% data reduction is achieved. (Unvoiced speech accounts for 30 to 40% of English language speech.) A second reduction is obtained by detecting periods during which the filter parameters may be the same as those of the previous frame. During these frames only the gain and pitch parameters are updated; this allows for an 80% data reduction.

In the AMI approach, the word-decode ROM, the speech-data ROM, and the coefficient ROM are mask programmed with the customer's speech data. The AMI provides a complete speech analysis service to support its speech synthesis activities. This allows accurate programming of the ROMs from a speech sample that is provided on audio magnetic tape. Customers with their own LPC speech analysis facilities may arrange for the coordination of quantization techniques and acceptable formats.

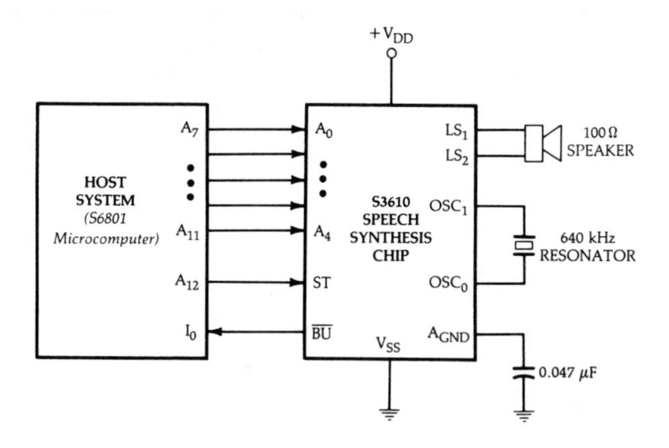

Fig. 9-2. Typical system configuration when an AMI S3610 synthesis chip is interfaced with an AMI S2000 microprocessor.

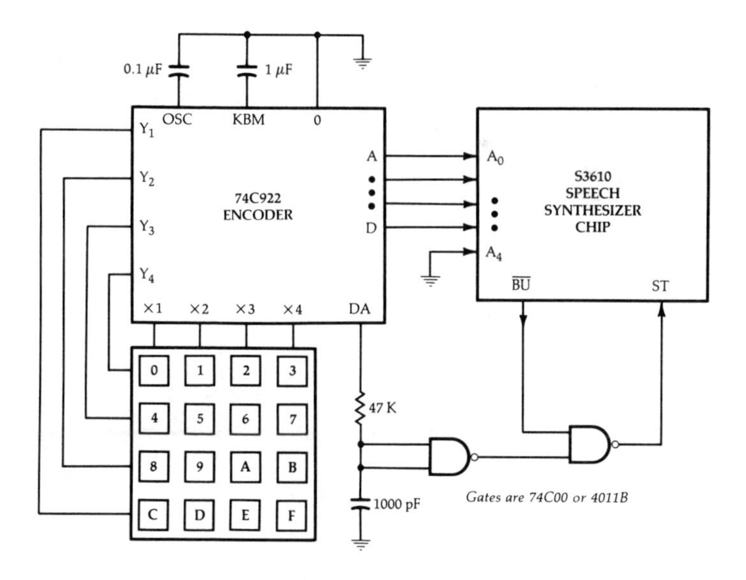

Fig. 9-3. Block diagram showing the direct keyboard operation of an AMI S3610 speech synthesizer chip.

The S3610 is designed so that it can be interfaced with a host controller but it may also be operated with a keypad using one or two encoder ICs. The S3620 is designed to interface with either 8-bit or 4-bit microprocessors that meet specified requirements. Fig. 9-2 is a typical system configuration that shows the S3610 synthesizer interfaced with an AMI S6801 microcomputer. Fig. 9-3 is a block diagram that illustrates the direct keyboard operation of the S3610 speech synthesizer chip.

CHAPTER 10

GENERAL INSTRUMENT PRODUCTS AND TECHNOLOGY

General Instrument Corporation (GI) has developed an effective voice synthesizer on a single silicon chip that will reduce the cost of adding the speech feature to various products. The chip employs formant coding, another type of frequency domain synthesis that is similar to linear predictive coding (LPC). The GI device also models speech as the output of a series of cascaded resonators that are alternately excited by switched periodic and aperiodic signals which correspond to voiced and unvoiced speech.

The data base for most formant synthesizers specifies only the center frequencies of the vocal tract resonances. The bandwidths of these frequencies are usually applied algorithmically, although they could be fixed values. A characteristic of formant coding is a lower requirement for digital memory than LPC because these frequencies need not be as well defined.

The all-pole LPC filter from General Instrument is designed so that the center frequency of each formant resonance may be adjusted independently of the related bandwidth. The benefits of this approach are increased flexibility in coding and, at the same time, a data base that is analogous to the one used in more conventional formant coding. The synthesizer functions like an LPC filter in operation.

The complete GI system consists of six cascaded resonators with each stage capable of simulating a single vocal tract resonance. The source is switchable between the digital signals representing voiced and unvoiced sources.

DEVICE DESCRIPTION

The General Instrument SP-0256 single-chip speech synthesis device contains all of the necessary functions for the performance of speech synthesis at a variable bit rate. The N-channel MOS device is configured in a single 28-pin DIP package. By itself, it is capable of synthesizing sixteen to thirty-five words without external memory. The block dia-

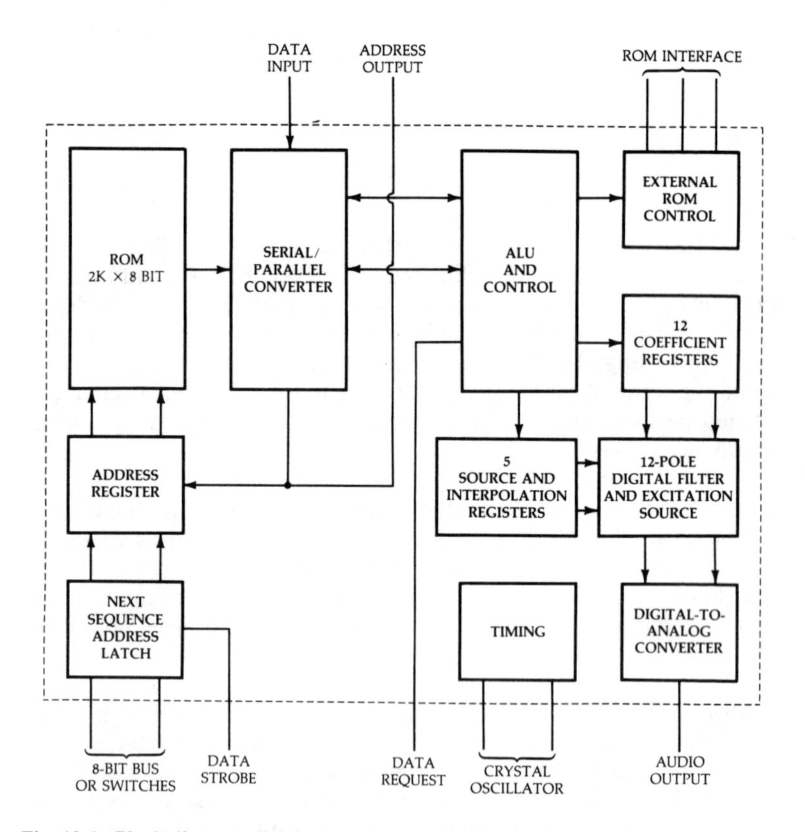

Fig. 10-1. Block diagram of a single-chip speech synthesis system from General Instrument Corp.

gram given in Fig. 10-1 shows that the SP-0256 chip has three main sections: the 16K-bit data and instruction ROM, the synthesizer, and the controller.

The heart of the system is the synthesizer with seventeen programmable parameters. It consists of the 12-pole digital filter and the logic for generating both a scaled pseudorandom noise source (unvoiced) and a periodic impulse source (voiced) with amplitude and period variable. The filter consists of two 16-by-10–bit serial-parallel multipliers and two serial adders. These are integrated into a single second-order stage. By multiplexing the stage six times with pipelining techniques, the response of a twelfth-order filter is obtained. A sampling rate of 10 kHz is applied.

The output of the filter section is connected to the on-chip digital-to-analog converter. By means of pulse-width modulation, the digital-to-analog converter generates a 7-bit quantized analog output. The analog output is applied to a low-pass filter, with a nominal cut-off frequency of 5 kHz, before it is amplified to drive the speaker.

The parameter registers of the synthesizer are updated by the 4-bit controller. The controller reads the input buffer to obtain the word or sound in the vocabulary and it carries out the instructions to make that sound. There are two basic kinds of SP-0256 instructions: register modification instructions to update the parameters controlling the synthesizer and individual instructions to modify the parameters singly, as a group, or completely. An instruction can either load immediate data from the speech ROM into a parameter register or add a two's complement value from the ROM to the existing parameter value. Each register modification instruction contains repeat bits which, in combination with the source parameters, indicate when the next modification will occur. Between register updates, the amplitude and period of the source are automatically smoothed under the control of the two interpolation registers.

Other instructions executed by the SP-0256 chip are those for branch control. They allow branches to any byte boundary in the 61K-byte address space. A single-level return-address stack is provided to allow commonly used words or phonetic units to be called as subroutines by a speech program elsewhere in memory. Return instructions are used to specify subroutine and speech program end points.

The SP-0256 speech synthesis chip interfaces to a series of custom speech ROMs for vocabulary expansion. Only seven lines are needed.

Communication is accomplished with the serial transfer of both address and data information. Each of the expansion ROMs in the system has an address counter that is incremented under the control of the SP-0256 chip. The 16-bit address is shifted out only when a jump instruction is executed. Address decoding is performed internally in the ROMs.

DEVICE APPLICATION

Application of the synthesizer to an existing host system is simplified because of the single 5-volt power supply, the crystal oscillator input, and the simple 8-bit port with two-line handshaking. The host system need only supply a vocabulary select code to the input port of the SP-0256 chip to synthesize the selected word, phrase, or sound. In host

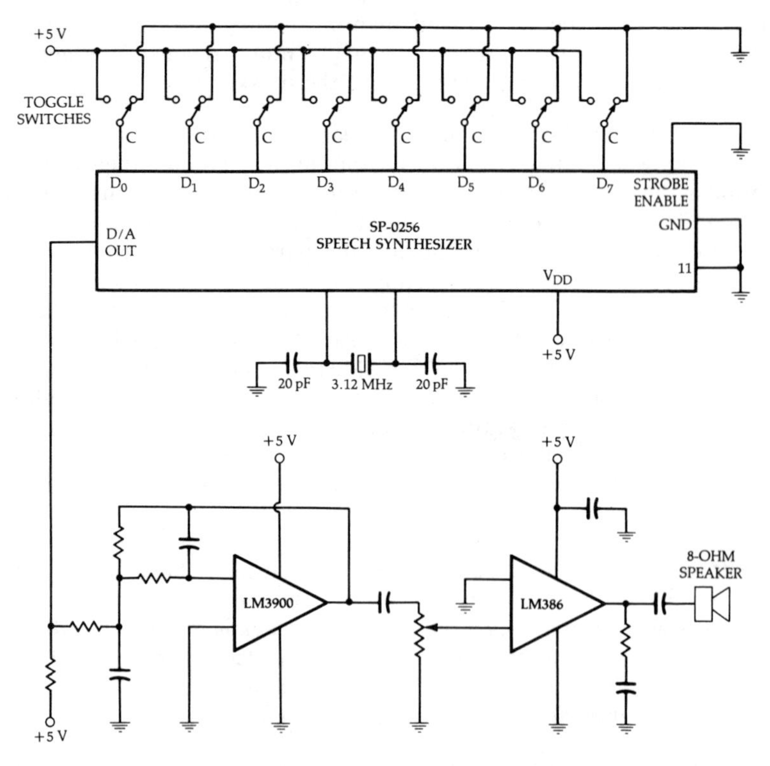

Fig. 10-2. Diagram showing the simple application of a General Instrument Corporation speech synthesis device.

systems where an input data strobe line is not available, the device may still be operated.

A power reduction capability is provided to conserve power when speech is not being synthesized. All inputs and outputs are TTL compatible.

A simple application for the SP-0256 chip is shown in Fig. 10-2. Closing any of the eight switches (D_0 through D_7) will cause one of eight phrases or words to be spoken. No strobe is necessary to load the data and switch debouncing is done internally. If the strobe enable pin is high, the 8-bit input word is strobed in using the ALD input. The configuration shown in Fig. 10-2 can also function as an independent annunciator system.

The SP-0256 chip can be interfaced with an 8-bit microcomputer chip and an SPR-16 vocabulary expansion ROM as shown in Fig. 10-3.

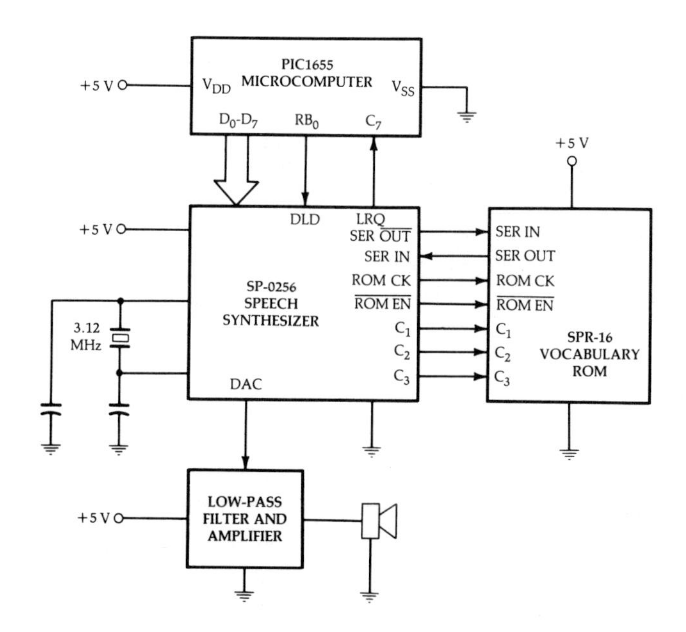

Fig. 10-3. Block diagram of a General Instrument Corp. SP-0256 single-chip speech synthesis device. The system configuration includes an 8-bit microcomputer interface and a 16K vocabulary expansion ROM.

General Instrument's PIC1655 microcomputer is used to control the eight input lines; the low-pass filter and amplifier are the same as shown in Fig. 10-2. The SPR-16 is a vocabulary expansion ROM from General Instrument. The diagram of Fig. 10-3 illustrates some typical interface requirements.

General Instrument Corp. is now offering three different vocabulary expansion ROMs: the 128K-bit SPR-128, the 32K-bit SPR-32, and the 16K-bit SPR-16. These ROMs may be paralleled externally in any combination necessary to match memory size with the specific vocabulary requirements of the end user.

CHAPTER 11

VOTRAX®
PRODUCTS AND TECHNOLOGY

The Votrax® Division of Federal Screw Works is one of the many companies offering applications-oriented speech synthesis devices and hardware. The products include Speech PAC™, which is a phoneme-access controller, the Versatile Speech Module (VSM/1™), and Type 'N Talk™, a text-to-speech synthesizer. All are based on the Votrax proprietary SC-01 speech synthesizer chip.

SC-01 SPEECH SYNTHESIZER

The Votrax SC-01 speech synthesizer is a completely self-contained speech synthesizer that is capable of phonetically synthesizing continuous speech, with unlimited vocabulary, from input data introduced at a low rate. Speech in the SC-01 device is synthesized by combining phonemes in an appropriate sequence to make meaningful speech. The device contains 64 different phonemes, which are accessible by a 6-bit code. The introduction of these phoneme codes to the SC-01 synthesizer, in the correct sequence, creates continuous speech. Signals from the SC-01 unit are applied to an audio output device for amplification and distribution of synthesized speech.

Speech PAC and VSM/1 are trademarks of Votrax®, a Division of Federal Screw Works.

The SC-01 speech synthesizer is an LSI circuit contained in a 22-pin DIP package. CMOS technology is used in the fabrication of the device to permit a high input impedance and a low power drain. The device is illustrated in Fig. 11-1. The SC-01 requires 70 bits of data per second to sustain continuous speech. The 6-bit phoneme codes are 5-volt logic compatible and are latched for data bus applications. A phoneme construction algorithm and filters within the chip create the synthesized audio output.

Fig. 11-1. The SC-01 speech synthesizer.

A block diagram of the SC-01 synthesizer is shown in Fig. 11-2. Data input is to six pins, P0 through P5, for the phoneme 6-bit selection code. Latching is controlled by the strobe (STB) signal. Latching occurs on the rising edge of the strobe signal. Inflection level setting is performed on pins 11 and 12 which instantaneously set the pitch level of voiced phonemes. The acknowledge/request (A/R) pin acknowledges receipt of phoneme data. The signal goes from high to low at one master clock cycle following the active edge of the STB signal. The A/R pin also indicates timing out of the old phonemes concurrent with a request for new phoneme data (signal goes from low to high). If external phoneme timing is desired, phoneme requests can be ignored.

The master clock resistor-capacitor (MCRC) input determines the internal master clock frequency. The RC values are selected for 720 kHz to obtain a standard phoneme timing. This point is connected to MCX when using an internal clock and is grounded when an external clock is used. Varying the clock frequency varies the voice and sound effects. As the clock frequency decreases, audio frequency decreases and phoneme timing lengthens.

The master clock external (MCX) allows control by an external clock signal. The MCRC is grounded during MCX operation. The audio output (AO) supplies the analog signal to the audio output device. The audio feedback (AF) pin is used with Class-A or Class-B transistor audio amplifiers for added stability. The Class-B (CB) pin is the current source for a Class-B transistor audio amplifier.

Table D-1, Appendix D, lists the 64 phonemes produced by the SC-01 synthesizer chip. Each phoneme code is accompanied by its symbol, average duration time, and an example. The underlined segments of the example word demonstrate the phoneme use, i.e., sound to be pronounced.

Table D-2, Appendix D, subdivides the 64 phoneme symbols into seven categories. Each category represents a different production feature. The first six categories are characterized by voiced fricative (expired voice) and nasal sounds. The seventh category is characterized by phonemes with no sound output.

PHONEME ACCESS CONTROLLER

The Votrax Speech PAC™ Phoneme Access Controller, based on the SC-01 speech synthesizer chip, provides the system designer with a small self-contained circuit board that is adaptable for use in industrial equipment, controllers, and games. The Speech PAC™ can be programmed by the customer and is expandable. The user can reconfigure the Speech PAC vocabulary, as desired. The EPROM socket may be jumpered to accept a 32K EPROM for stored vocabulary expansion. Phonemes and prestored words can be mixed as desired to produce an output of unlimited vocabulary. The Speech PAC™ is illustrated in Fig. 11-3.

The Speech PAC™ offers parallel interface for computer, controller, or preselected diode matrix to permit access to prestored words or to create phonetic speech. The on-board audio amplifier has a volume control.

Fig. 11-4 is a block diagram of the Speech PAC™ in the prestored word mode, while Fig. 11-5 is a block diagram of the Speech PAC™ in the phoneme mode.

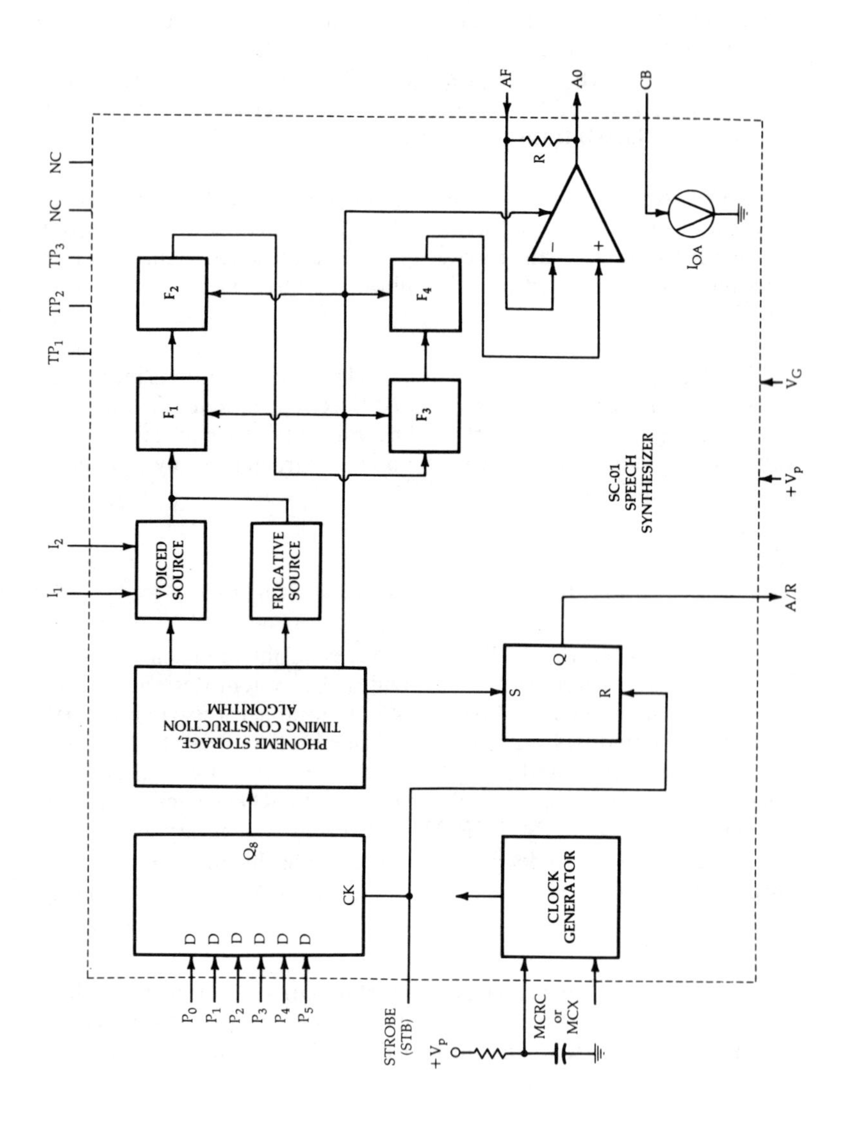

Fig. 11-2. Block diagram of the Votrax® SC-01 speech synthesizer chip.

Fig. 11-3. An isometric photo of the Vortrax® Speech PAC™ Phoneme Access Controller.

Prestored words are accessed in 8-byte increments. The low baud rate of the SC-01 speech synthesizer allows a single 2716 EPROM to store up to 255 words and a single 2532 EPROM to store up to 511 words. Phoneme sequences that are more than eight phonemes long may cross entry boundaries. The Speech PAC™ signals the external controller at the end of each phoneme sequence.

VERSATILE SPEECH MODULE

The Votrax Versatile Speech Module (VSM/1™) is a board-level speech synthesis module intended for use as a microcomputer in the simulation or development of products with synthesized speech. It can also be used for real-time speech synthesis while simultaneously executing commands and performing monitoring activities. It is designed to be plugged directly into the card cage of an industrial control computer to provide instructions for employees performing tasks in real time. This application may be of use in chemical processing plants, nuclear power stations, aircraft systems, seismic monitoring stations, and in automated warehouses. The VSM/1™ Versatile Speech Module is shown in Fig. 11-6.

The VSM/1™ can operate in a computer-to-computer mode for task downloading, in addition to voice synthesis. It can also be used in a distributed processing mode. Instructions and activities can be dynami-

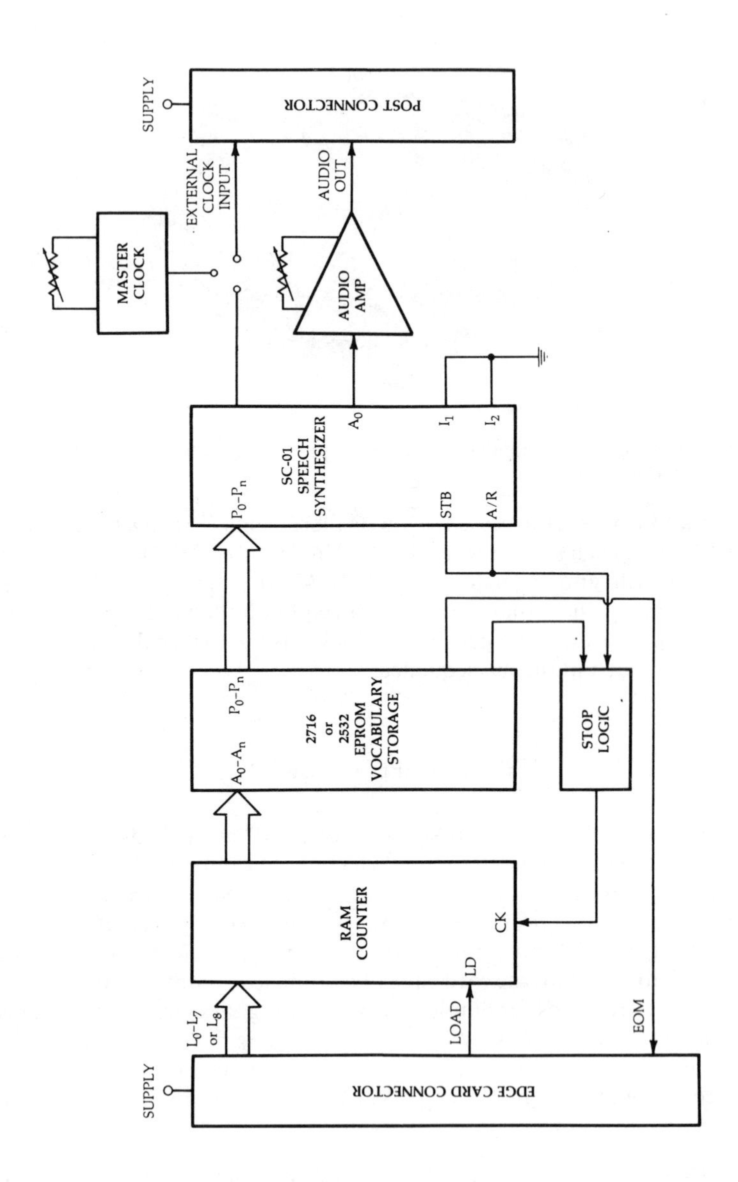

Fig. 11-4. Block diagram of the Votrax® Speech PAC™ phoneme access controller in the prestored word mode.

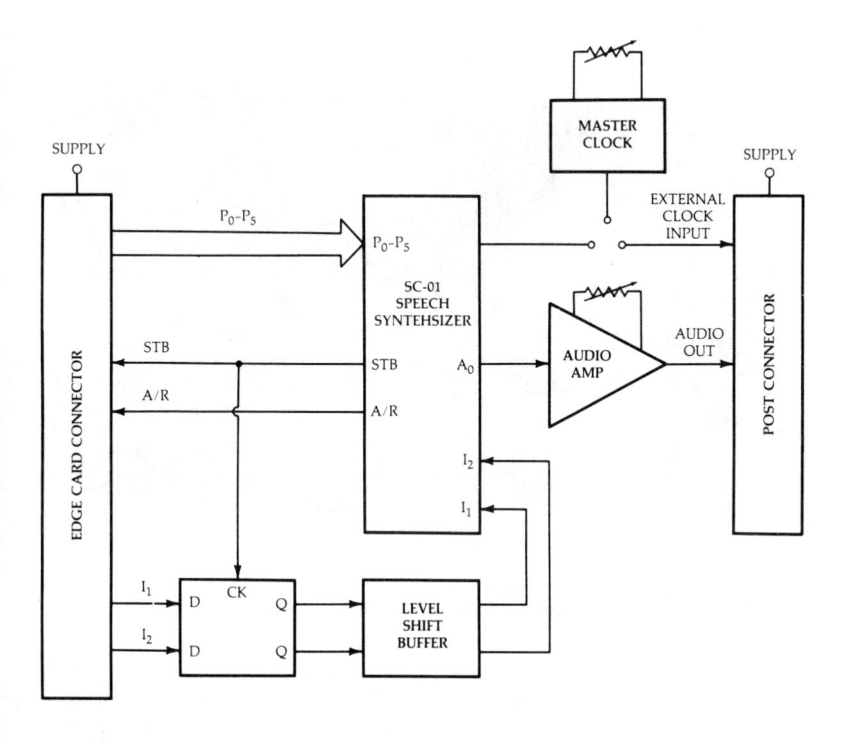

Fig. 11-5. Block diagram of the Votrax® Speech PAC™ phoneme access controller in the phoneme mode.

cally supplied from another computer or from a user ROM. Fig. 11-17 is a block diagram of the VSM/1™.

Speech rate, pitch, and pause controls can be dynamically programmed, by control codes, to produce various stress patterns. By altering the master clock controls, many human voice effects can be programmed to simulate a multivoiced environment. The VSM/1™ can produce a multitude of sound effects from prestored sound macros. Space sounds, gunfire, explosions, race cars, and even musical sequences can be created. Additional sounds can be user defined and stored in the EPROMs.

The voice operating system (voxOS) may be bypassed, if desired. The user can supply the necessary firmware to change the system function. This may be done in changing from an experimental system to actual application hardware.

Fig. 11-6. The VSM/1™ Versatile Speech Module by Vortrax®.

The VSM/1™ has four audio-sequence memories for loading speech callout codes, system controls, prestored speech, and/or sound effects. System changes can be accomplished by placing a user-supplied EPROM in one of the vacant card sockets on the speech module. The VSM/1™ can be used as a microcomputer to talk and execute commands, or as a smart peripheral that is expandable by interface ports. System functions can be changed by downloading a 6800-compatible code segment.

The versatile speech module includes, in addition to the SC-01 phoneme synthesizer, a 6800 microprocessor unit, and a parallel and serial (RS-232) interface that has a selectable baud rate of 75–9600 bits per second. The module includes a 1K-byte RAM with sockets for an additional 2K bytes, a 2K-byte voxOS operating system, an 8K-byte prestored vocabulary ROM, and expansion sockets for an additional 8K-bytes (device 2716) to 16K bytes (device 2532) of jumper-selectable EPROMs.

The 8-ohm 1-watt on-board amplifier has a volume control. A half-memory plane expansion connector provides 32K locations out of 64. Customer access to the 32K locations is by means of the microcomputer data address bus.

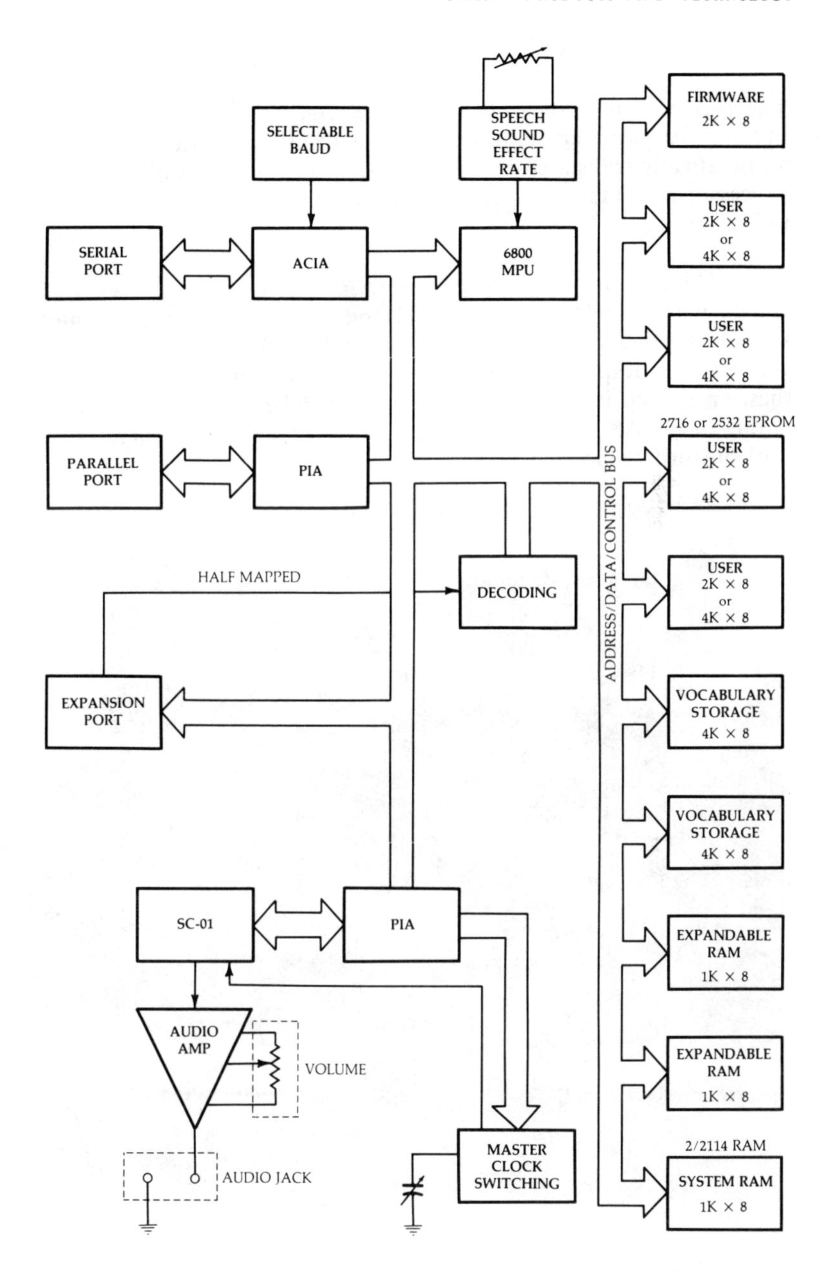

Fig. 11-7. Data flow diagram of the Votrax® VSM/1™ versatile speech module.

TEXT-TO-SPEECH SYNTHESIZER

The Type'N Talk™ text-to-speech synthesizer from Votrax® is an item of computer peripheral equipment that permits typewritten words to be automatically translated into electronic speech by the system's microprocessor-based text-to-speech algorithm. The Type'N Talk™ is shown in Fig. 11-8.

English text is automatically translated into electronically synthesized speech with Type'N Talk™. ASCII code from the host computer's keyboard is fed to Type'N Talk™ through an RS-232C interface to generate synthesized speech. The English word is entered and the synthesized speech is heard from the audio loudspeaker. For example, typing in the ASCII characters representing "h-e-l-l-o" generates the spoken word "hello."

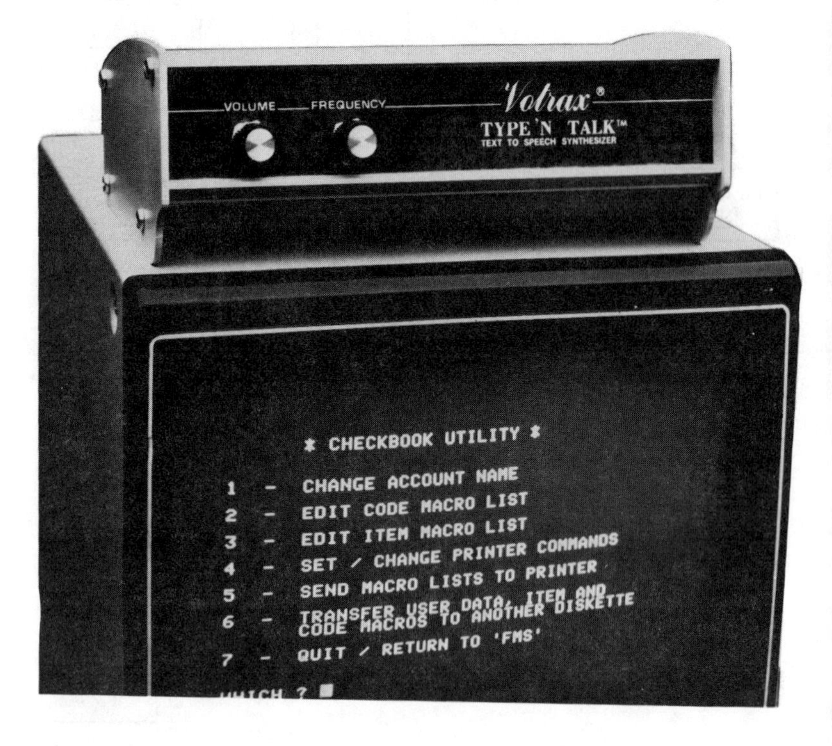

Fig. 11-8. The Type'N Talk™ text-to-speech synthesizer from Votrax®.

Type'N Talk™ has a built-in microprocessor and a 750-character buffer to hold typed words. Desk-top computers can execute programs and speak simultaneously. The unit does not require the use of the host computer's memory. Neither does it involve the computer with time consuming data translation.

The Type'N Talk™ synthesizer is located between a computer or modem and a terminal. It can speak all the data that's sent to the terminal while on-line with a computer. Information randomly accessed from a data base can be verbalized. With the unit's data switching capability, it can be "de-selected" while data are sent to the terminal and vice-versa—permitting speech and visual data to be independently sent on a single data channel.

The Type'N Talk™ text-to-speech synthesizer can be interfaced in several ways using special control characters. For example, it can be connected directly to a computer's serial interface. Then, a terminal, a line printer, or an additional Type'N Talk™ unit can be connected to the first unit, eliminating the need for additional RS-232C ports on the host computer. If the user employs unit assignment codes, multiple Type'N Talk™ units can be daisy-chained. Unit addressing codes allow the Type'N Talk™ units and the host printer to be independently controlled.

CHAPTER 12

TELESENSORY SPEECH SYSTEMS PRODUCTS AND TECHNOLOGY

Telesensory Speech Systems of Palo Alto, CA, is a participant in speech synthesis at the functional assembly level. They offer pre-engineered solutions to end users and to OEMs so that the complexities of circuit integration can be avoided. The board-level products are intended for use in actual applications rather than as demonstrators. They may be tailored to specific tasks through the selection of vocabulary or by means of device configuration on the pc board.

SPEECH SYNTHESIZER MODULE

The Series III Speech Synthesizer Module from Telesensory Speech Systems is a complete speech synthesizer circuit board with all the components necessary for installation in a computer-based end product. It can accommodate both standard and custom vocabulary memories. It has a capacity of 256 voice sounds—typically letters, numbers, or commonly used words. This permits 100 seconds of synthesized human speech.

The block diagram of the Series III Module is shown in Fig. 12-1. Telesensory Speech Systems says that it has applications in automated

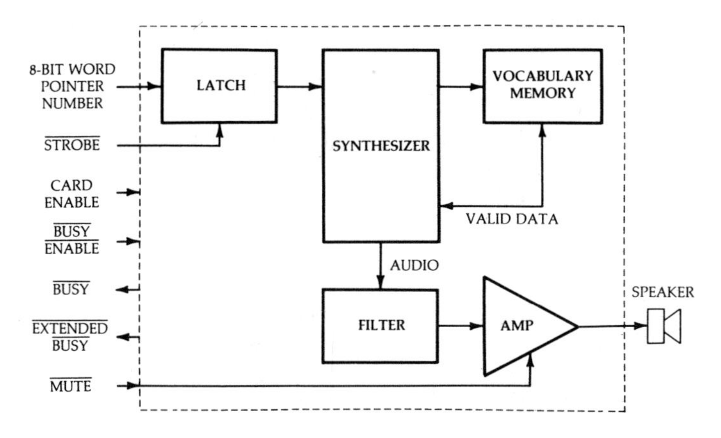

Fig. 12-1. Block diagram of the Series III speech synthesizer module from Telesensory Speech Systems.

test equipment, instrumentation, alarm and security systems, and telecommunications. The module includes Telesensory's proprietary time-domain encoded synthesizer device, a 119-word basic vocabulary, a speech filter and amplifier, a microcomputer interface, and an extra ROM socket. These are all mounted on a small circuit board that measures 4 inches by 4½ inches (10.16 cm x 11.43 cm). Interfacing, by means of standard card edge connectors and flat cable headers, can be accomplished to most popular microprocessor buses, TTL-compatible I/O ports, or simple logic controllers. The power requirement is a single 5-volt source. A speaker must be provided.

MODULE OPERATION

Speech is initiated with an 8-bit parallel digital code that defines the word or phrases to be spoken and which includes a start signal. The synthesizer chip obtains compressed speech data from the vocabulary memory and reconstructs the speech waveform as an analog audio signal output by means of the on-chip digital-to-analog converter. The audio signal is filtered and amplified on the module to produce a male voice.

The speech module includes a set of basic vocabulary words that are stored in a 64K ROM. If those words are insufficient, they can be augmented with special words or phrases in 16K-, 32K-, or 64K-bit memories. The vocabulary word list is given in Appendix E.

Phrases can be enunciated by concatenating words in the Series III as follows: /in/ /-crease/, /out/put/, /rate/, /two/, /fif-/ teen/, /liters/, /per/, /hour/. The appropriate amount of silence between words will depend on both the words used and the effect desired. The sounds /s/ and /z/ can be used to convert singular words to their plural form. Other words can be created by using the basic vocabulary set as follows: /B/ /4/ before; /Y/ why; /U/ you; /re-/ /cent/ recent; /O/ /K/ okay; etc.

Telesensory Speech Systems offers its customers a choice of words in a lexicon list and it encodes them on EPROM or ROM. The company also offers a custom service and will encode words on EPROM or ROM as well. The unused socket on the Series III speech module is available for custom vocabularies.

SPEECH 1000 SYNTHESIZER BOARD

In moving to yet higher levels of integration with modular speech synthesis boards, Telesensory Speech Systems introduced the Speech 1000 synthesizer board. Terming it a stored vocabulary speech response unit, the Speech 1000 board contains an 8085A microprocessor, a control program in ROM, a proprietary digital signal processor, a 10-bit digital-to-analog converter, a 7-pole anti-aliasing low-pass filter, a 2-watt audio amplifier, three alternative interface I/O ports, and seven memory sockets for vocabulary storage.

The block diagram for the Speech 1000 board is shown as Fig. 12-2. The synthesizer board is controlled by the host computer through commands and word pointers. Commands instruct the board to stop, repeat, or increase the speaking rate. Word pointers specify the speech element to be spoken. By arranging the commands and word pointers in a desired sequence and then downloading them into a RAM buffer on the Speech 1000 board, the host system's need for polling or interrupt handling is reduced.

Once the command and word pointers are loaded into the Speech 1000 board, the microprocessor retrieves the necessary encoded parametric speech data from the vocabulary memory, processes it, and presents it to the signal processor for synthesis.

Unlike the Series III module, the Speech 1000 module employs a linear predictive coding (LPC) approach. The processor is structured as a

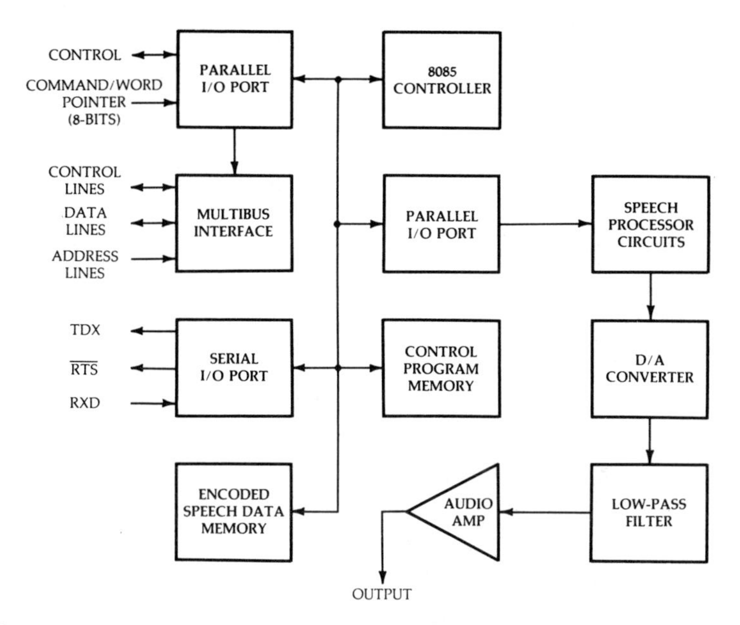

Fig. 12-2. Block diagram of the programmable digital signal processor (PDSP) from Telesensory Speech Systems.

12-pole lattice filter. It operates at the rate of 10,000 10-bit samples per second.

The vocabulary memory can have as many as 458K bits but it may be as small as 16K bits. Memory sockets will accommodate 16K, 32K, or 64K ROMs, or 32K or 64K EPROMs.

The digital-to-analog converter transforms the digital representation into its analog counterpart for amplification. The "on-board" amplifier supplies up to two watts into an 8-ohm speaker. Volume control can be accomplished either by the "on-board" potentiometer, by connecting an external potentiometer, or by using the down-loadable software command "attenuate amplitude."

The Speech 1000 board can synthesize multiple voices, either male or female. The Speech 1000 module is a pc board with the Intel MULTIBUS® form factor. It measures 6.75 inches by 12.00 inches and is 0.5 inches high (17.15 cm x 30.48 cm x 12.7 cm).

APPENDIX A

GLOSSARY OF ELECTRONIC SPEECH SYNTHESIS TERMS

Allophone—A variation of a phoneme as conditioned by its position or by adjoining sounds. Example: The short (a) of mat and the slightly longer (a) of mad are allophones. It is considered that there are 128 allophones as compared to about 40 phonemes in the English language. Allophone use improves the quality of speech synthesis.

Analog-to-digital conversion—The process of converting an analog or time-varying signal such as the output of a telephone to a coded digital signal representation of that analog signal. This is often done to permit the monitoring or processing of data by a computer.

Articulation—The way in which parts of speech are joined together. Also, a way of talking or pronunciation.

Band-pass filter—A filter with a single transmission band that permits a restricted part of the frequency spectrum to pass. The filter attenuates or suppresses frequencies on either side of this band.

Binary digit—Usually abbreviated as bit. A unit of information equal to one binary decision. The designation of one of two possible states as either a 1 or a 0. Thus, a bit is said to have the value of either ONE (1) or ZERO (0).

Complementary metal-oxide semiconductor (CMOS)—A technique for the manufacture of semiconductor devices that is most important in the fabrication of large-scale integrated (LSI) circuits. A variation of the metal-oxide semiconductor (MOS) p-channel and n-channel processes. CMOS permits low power consumption and is tolerant of slight variations in the power supply.

Compressed speech—A term used to describe digitized speech with the redundant features removed to reduce the data content for more economical transmission and storage.

Concatenation—A linking together of a series of things or events. In speech synthesis, it is the linking together of words or sounds in memory to form phrases or sentences.

Consonant—Any speech sound produced by stopping and releasing the air stream (p, t, k, b, d, g), or by stopping it at one point in the vocal tract while it escapes at another (m, n, l, r), or by forcing it through a loosely closed or very narrow passage (f, v, s, z, sh, zh, th, h, w, y), or by a combination of these means (ch, j). *See also* vowel and fricative.

Constructive synthesis—A speech synthesis technique that "constructs" speech sounds from a library of elemental sounds. This term contrasts with the expression "analysis synthesis" that refers to methods based on the recording of human speech and the subsequent processing of it into a digital format by various techniques, including LPC and formant synthesis.

Continuous variable-slope delta modulation—An extension of the delta modulation technique that allows the amplitude change (slope) to vary.

Data rate—The data transmission capacity, expressed in bits per second, required to convey a given digitizing process. It is also an indicator of the memory needed to store discrete amounts of data representing a specified activity. In speech synthesis, this would mean the amount of memory required to store a word or sentence in a digitized form.

Delta modulation—A type of waveform encoding that encodes the differences between individual speech samples. This technique permits the determination of the next data-point value by examining the changes in amplitude of the existing data point.

Digital-to-analog conversion—The process of converting a digital representation of an analog or time-varying signal to an analog signal. It is the opposite of analog-to-digital conversion.

Digitized speech—The result of analog-to-digital conversion of the speech waveform as a result of sampling at regular intervals. The accuracy of the representation depends on the sampling rate and the length of the digital word in bits. Speech is typically sampled at rates from 8,000 to 12,500 samples per second.

EEPROM or **E^{2}PROM**—Abbreviation for Electrically Erasable Programmable Read-Only Memory. This programmable read-only memory, like the EPROM, is manufactured without a data pattern. The user writes in specified data by means of special programming equipment that makes nondestructive semipermanent changes in the memory elements, EEPROMs can be reprogrammed following an erasure, which is performed by the application of electrical overvoltages. These overvoltages permit selective or complete erasure. *See also* EPROM, PROM, and ROM.

EPROM—Abbreviation for Erasable Programmable Read-Only Memory. This programmable read-only memory, like the EEPROM, is manufactured without a data pattern. The user writes in specified data by means of special programming equipment makes nondestructive semipermanent changes in the memory elements. EPROMs can be reprogrammed following their erasure. Erasure is usually performed by exposing the entire memory chip to ultraviolet light. *See also* EEPROM, PROM, and ROM.

Excitation—The driving signal or signals in any speech models. In a direct mechanical analogy, excitation is by means of compressed air; in a digital analogy, it is performed by periodic and aperiodic digital signals that represent the movement of vocal cords or other constraints in the vocal tract.

Formant—Formants are resonances in the frequency spectra of voiced speech. They appear as bands or dense concentrations in the spectrographic display of voiced speech. Formants are harmonic resonances that help to distinguish sounds.

Formant synthesis—A parameter encoding technique that models speech information by tracking formants of the frequency spectrum.

Frame—A segment of time in an utterance. A single frame usually represents 10 to 25 milliseconds of speech. *See also* linear predictive coding.

Frequency—The number of recurrences or cycles of a periodic phenomenon in a specified unit of time.

Fricative sounds—Speech sounds that are produced when air is expelled from the lungs through a constricted vocal tract to produce turbulent noises. The "s" and "ch" sounds in the word *speech* are two examples. *See also* voiced and unvoiced sounds.

Linear predictive coding—Abbreviated LPC. A parametric encoding technique that models the human voice tract with a digital filter whose controlling parameters change with respect to time. Each set of controlling parameters is called a *frame*.

Metal-oxide semiconductor (MOS)—A process for the manufacture of semiconductors that is particularly important in fabricating integrated circuits. The two prominent forms, p-channel and n-channel, are used to fabricate dense large-scale integrated circuits. Some examples include microprocessors, memories, and speech synthesizers.

Microcomputer—A development in microprocessors that incorporates a functional memory, input/output functions, and control functions on a single integrated-circuit chip along with the microprocessor or CPU function. Microcomputer chips are sometimes referred to as microcontrollers.

Microprocessor—A computer central processing unit (CPU) that is contained on a single integrated-circuit chip. To function as a computer, a microprocessor needs ancillary or peripheral integrated circuits to perform input/output, memory, and, sometimes, control functions.

Parameter—A quantity or constant whose value varies with the circumstances of its application. It is a constant with variable values that is used in reference to other variables.

Parametric speech encoding—An encoding technique that represents specific properties of speech (e.g., pitch, energy, etc.) to describe the unique characteristics of both the speaker and the language.

Phonemes—A set of the simplest and most flexible basic sounds present in a language. They are the building blocks of speech from which all speech sounds may be synthesized. The English language has about 40 phonemes; 16 are vowel phonemes and 24 are consonant phonemes. Since there may be many minor variations in the actual pronunciation of each phoneme, another linguistic variation called the *allophone* is used to convey the phonemes more precisely.

Phonetics—The branch of language study that is concerned with speech sounds, their production, and their combination. Special symbols are used that represent words and phrases graphically, or phonetically.

Pitch—The term used to indicate frequency in audible sound—voice, sound effects and music. In speech, pitch is determined by the vibration frequency of the vocal cords. When pitch is raised one octave, the frequency is doubled.

Plosive sounds—Speech sounds that are created when the vocal cavity is constricted by the lips and tongue, thus permitting an air pressure buildup. The release of air when expressing the sounds "d" and "p" is two examples. *See also* unvoiced sounds and voiced sounds.

PROM—Abbreviation for Programmable Read-Only Memory. This read-only memory, like the EEPROM and EPROM, is manufactured without a data pattern. The user writes in specified data by means of special programming equipment that makes destructive permanent changes in the memory elements. PROMs cannot be reprogrammed. *See also* EEPROM, EPROM, and Read-Only Memory (ROM).

Prosody—A term that refers to tone or accent as in a song sung to music. In linguistics, it refers to the energy, stress, intonation, pause, and duration characteristics of speech that distinguish it from a flat monotonic expression.

Pulse code modulation (PCM)—A technique for sampling a waveform at a set rate, say 8000 times per second, and, then, coding and storing information about the amplitude in a continuous process. The signal is restored by digital-to-analog conversion.

Random-Access Memory (RAM)—A type of memory that can be accessed at random as opposed to serial memory (a shift register or a charge-coupled device), that requires finite playback in order to retrieve a passage or section of data. As commonly used today, RAM refers to a volatile high-speed memory that loses its stored data if the power is interrupted. There are two forms of RAM—static and dynamic. A dynamic RAM must be periodically refreshed in order to retain the stored data. It is often referred to by the designation DRAM—for dynamic RAM. At the time of this writing, the largest commercially available DRAMs contain 65,536 bits, but they are conventionally referred to as 64K DRAM. All RAMs are termed read-write memories.

Read-Only Memory (ROM)—A read-only memory is a nonvolatile memory that may not be erased or rewritten. Data patterns are permanently masked into the device memory during manufacture. ROMs are randomly accessed, the same as RAMs. They are also referred to as semicustom memories. *See also* EEPROM, EPROM, and PROM.

Sampling rate—The number of samples, taken per second, of a value or waveform and used to represent a signal. In general, the higher the sampling rate, the higher the resolution, but the larger the amount of data that are required.

Spectrogram—A presentation of sound or speech in which the component frequencies and the intensities of those frequencies are represented. Typically, time is represented graphically on the horizontal or X-axis and frequency that is represented on the vertical or Y-axis. Intensity is represented by the gradations of darkness of the graphics.

Spectrum—The distribution of frequencies in a waveform. The presentation on a spectrum analyzer typically shows signal amplitude as a function of frequency that is presented in ascending order from left to right on the horizontal axis.

Stress—The accenting of the human voice so that certain syllables sound louder than others.

Synthetic speech—Artificially reproduced sounds that are interpreted by the human ear as human speech. Although synthesis can be performed mechanically, today it is most widely effected by the use of digital electronic techniques.

Text-to-speech—A method for producing synthetic speech, with the aid of a computer program, that accepts input in the form of typed words.

Unvoiced sounds—Speech sounds produced without the periodic vibrations of the vocal cords. These may be further classed as *plosive* and *fricative* sounds.

Voiced sounds—Sounds produced by the periodic sound-wave generating vibrations of the human vocal cords. Voiced sounds, as well as unvoiced sounds, may be either fricative or plosive.

Vowel sounds—A voiced speech sound that is characterized by a general restraint on the air which is passing in a continuous stream through the pharynx and the opened mouth, but with no constriction narrow enough to produce local friction. Vowels are the sounds of greatest prominence in most syllables. The letters *a, e, i, o, u* and, sometimes, *y* represent vowels sounds.

APPENDIX B

TEXAS INSTRUMENTS' SPEECH DEVELOPMENT SYSTEM

The following lists cover the equipment required or recommended for the Speech Development System (SDS) from Texas Instruments Incorporated.

REQUIRED EQUIPMENT

The DS990 Minicomputer provides the host computing environment for all SDS packages. The recommended system configuration for product operation of the SDS package is a Multi-AMPLUS™ Development System (TMAM9080-XX). The "-XX" notation identifies which audition board is included in the system.

Item	Part No.	Description
Multi-AMPLUS™ Development System	TMAM9080-XX	990/12 CPU, 50 MB disk and drive, 1600 bpi magnetic tape drive, 256 KB core

		memory, 3 KB cache memory, two Model 991 VDTs, Multi-AMPLUS (DX10 operating system, diagnostics, utilities, PROM software), a VSP audition board, and the 16-bit I/O card.
Data Collection Processor	SPSH1001	Remote analog interface for analyzing, recording, and digitizing speech.
DCP-DS990 Interface		
a. Parallel cable	SPSH1102	50-foot I/F cable.
b. Analog cable	SPSH1103	Two 6-foot I/F cables.
VSP Audition Board (-XX)		
a. 5100 board (-01)	TMAM6078	TMS5100 VSP board.
b. 5200 board (-03)	TMAM6079	TMS5200 VSP board.
c. 5220 board (-04)	TMAM6082	TMS5220 VSP board.
Speech Composition and Analysis Software Package	SPSS2001-08	DS990 SDS software including SCS editor, script creator, speech linker, recording session control, and LPC utilities.
PROM Programmer Kit	TMAM6022	PROM/EPROM master kit.
PROM Programmer Card		
a. 27XX series adapter	TMAM6024	3-supply adapter.
b. 25XX series adapter	TMAM6056	Single 5-volt adapter.
c. PROM II (TB2) adapter	TMAM6055	Bipolar PROM adapter.

d. EPROM erase adapter	TMAM6025	EPROM erase kit.

RECOMMENDED SYSTEM OPTIONS

To provide maximum performance and flexibility to the system, the following items are offered as options by Texas Instruments Incorporated.

Item	Part No.	Description
810 Printer	TMAM7001	120-cps printer kit.
LP300 Printer	TMAM7002	300-lpm printer kit.
50-MB Expansion Memory	TMAM2042	DS50 disk memory.
256-KB Expansion RAM	TMAM2083	Expansion core memory.
Expansion RAM Cables	TMAM2002	Expansion core cable kit.
Dual 911 VDTs	TMAM7002	Expansion VDT master kit.
Recommended Memories		Light use—50-MB disk, medium use—100-MB disk, heavy use—200-MB disk. (Expansion core memory is recommended for faster access times and for additional computational purposes.)
Other Options		An additional DCP is required for each additional sound recording operation.

APPENDIX C

VOCABULARY LISTS

The following lists contain the word, sound, and letter contents of several factory mask-programmed vocabulary read-only memories (VROMs) from Texas Instruments Incorporated.

The VM71001 ROM contains a 50-item industrial vocabulary that is stored in a 32K VROM. It is compatible with the TMS5100 voice synthesis processor. The sounds are spoken in a male voice. Next is the vocabulary for the VM71002, spoken in a male voice, and the VM71003, spoken in a female voice. They have a 35-item numeric/time vocabulary that is contained in a 32K VROM. The VM71002 is compatible with the TMS5100 VSP while the VM71003 is compatible with the TMS5220 processor. Finally, there is a 206-item industrial vocabulary that is stored in a 128K VROM and spoken in a male voice. This is the vocabulary for a VM61002 ROM; it is compatible with the TMS5100 voice synthesis processor.

VM71001 VOCABULARY

ZERO	TWO	FOUR	SIX
ONE	THREE	FIVE	SEVEN

EIGHT	HUNDRED	FARAD	COMPLETE
NINE	THOUSAND	WATTS	CONNECT
TEN	A	METER	DEGREES
START	M	OHMS	MINUS
IS	P	AREA	REPAIR
ELEVEN	T	LIGHT	SECONDS
TWELVE	STOP	PRESSURE	SERVICE
THIR-	AND	OFF	NOT
FIF-	THE	POWER	TEMPERATURE
-TEEN	AMPS	CHECK	ON
TWENTY	HERTZ		

VM71002 and VM71003 VOCABULARIES

ONE	SEVEN	THIRTEEN	NINETEEN
TWO	EIGHT	FOURTEEN	TWENTY
THREE	NINE	FIFTEEN	THIRTY
FOUR	TEN	SIXTEEN	FORTY
FIVE	ELEVEN	SEVENTEEN	FIFTY
SIX	TWELVE	EIGHTEEN	A.M.
THE	TIME	IS	P.M.
GOOD	MORNING	EVENING	AFTERNOON
O'CLOCK	OH	(Slight pause)	

Note that one of the words for these ROMs is not an actual sound but is, instead, a *slight pause*.

VM61002 VOCABULARY

ZERO	TWELVE	F	R
ONE	THIR-	G	S
TWO	FIF-	H	T
THREE	-TEEN	I	U
FOUR	TWENTY	J	V
FIVE	HUNDRED	K	W
SIX	THOUSAND	L	X
SEVEN	A	M	Y
EIGHT	B	N	Z
NINE	C	O	ALPHA
TEN	D	P	BRAVO
ELEVEN	E	Q	CHARLIE

DELTA
ECHO
FOXTROT
OPERATOR
GOLF
HENRY
INDIA
JULIET
KILO
LIMA
MIKE
NOVEMBER
OSCAR
PAPA
QUEBEC
ROMEO
SIERRA
TANGO
UNIFORM
VICTOR
WHISKEY
X-RAY
YANKEE
ZULU
AND
THE
AMPS
HERTZ
FARAD
WATTS
MEGA
MICRO
MILLI
METER
PICO
OHMS
CAUTION
DANGER
FIRE
AREA
LIGHT
PRESSURE
POWER
CIRCUIT
CHECK
CHANGE
COMPLETE
CONNECT
DEGREES
MINUS
REPAIR
SECONDS
SERVICE
NOT
TEMPERATURE
GALLONS
UNIT
SWITCH
START
STOP
TIMER
VALVE
LINE
MACHINE
UP
DOWN
OFF
ON
IS
NUMBER
TIME
CONTROL
ALERT
OUT
AUTOMATIC
ELECTRICIAN
ADJUST
POINT
WAIT
AT
BETWEEN
BREAK
SMOKE
RED
MINUTES
HOURS
ABORT
ALL
BUTTON
CALIBRATE
CALL
CANCEL
CLOCK
CRANE
CYCLE
DAYS
DEVICE
DIRECTION
DISPLAY
DOOR
EAST
ENTER
EQUAL
EXIT
FAIL
FEET
FAST
FLOW
FREQUENCY
FROM
ABOUT
GAGE
GATE
GET
GO
GREEN
HIGH
HOLD
INCH
INSPECTOR
INTRUDER
LEFT
LOW
MANUAL
MEASURE
MILL
MOTOR
MOVE
NORTH
OF
OPEN
OVER
PASS
PASSED
PERCENT
PLUS
POSITION
PRESS
PROBE
PULL
PUSH
RANGE
READY
REPEAT
RIGHT
SAFE
SET
SHUT
SLOW
SOUTH
SPEED
TEST
TOOL
TURN
UNDER
VOLTS
WEST
YELLOW

APPENDIX D

VOTRAX® SC-01 SPEECH SYNTHESIZER VOCABULARY

Table D-1. Phoneme Chart

Phoneme Code	Phoneme Symbol	Duration (ms)	Example Word
ØØ	EH3	59	jacket
Ø1	EH2	71	enlist
Ø2	EH1	121	heavy
Ø3	PAØ	47	no sound
Ø4	DT	47	butter
Ø5	A2	71	made
Ø6	A1	103	made
Ø7	ZH	90	azure
Ø8	AH2	71	honest
Ø9	I3	55	inhibit
ØA	I2	80	inhibit

Table D-1—cont. Phoneme Chart

Phoneme Code	Phoneme Symbol	Duration (ms)	Example Word
0B	I1	121	inhibit
0C	M	103	mat
0D	N	80	sun
0E	B	71	bag
0F	V	71	van
10	CH*	71	chip
11	SH	121	shop
12	Z	71	zoo
13	AW1	146	lawful
14	NG	121	thing
15	AH1	146	father
16	OO1	103	looking
17	OO	185	book
18	L	103	land
19	K	80	trick
1A	J*	47	judge
1B	H	71	hello
1C	G	71	get
1D	F	103	fast
1E	D	55	paid
1F	S	90	pass
20	A	185	day
21	AY	65	day
22	Y1	80	yard
23	UH3	47	mission
24	AH	250	mop
25	P	103	past
26	O	185	cold
27	I	185	pin

Table D-1—cont. Phoneme Chart

Phoneme Code	Phoneme Symbol	Duration (ms)	Example Word
28	U	185	move
29	Y	103	any
2A	T	71	tap
2B	R	90	red
2C	E	185	meet
2D	W	80	win
2E	AE	185	dad
2F	AE1	103	after
30	AW2	90	salty
31	UH2	71	about
32	UH1	103	uncle
33	UH	185	cup
34	O2	80	for
35	O1	121	aboard
36	IU	59	you
37	U1	90	you
38	THV	80	the
39	TH	71	thin
3A	ER	146	bird
3B	EH	185	get
3C	E1	121	be
3D	AW	250	call
3E	PA1	185	no sound
3F	STOP	47	no sound

/T/ must precede /CH/ to produce CH sound.
/D/ must precede /J/ to produce J sound.

Table D-2. Phoneme Categories According to Production Features

Voiced		"Voiced" Fricative	"Voiced" Stop	Fricative Stop	Fricative	Nasal	No Sound
E	EH	Z	B	T	S	M	PAØ
E1	EH1	ZH	D	DT	SH	N	AP1
Y	EH2	J	G	K	CH	NG	STOP
Y1	EH3	V		P	TH		
I	A	THV			F		
I1	A1				H		
I2	A2						
I3	AY						
AE	UH						
AE1	UH1						
AH	UH2						
AH1	UH3						
AH2	O						
AW	O1						
AW1	O2						
AW2	OO						
OO1	IU						
R	U						
ER	U1						
L	W						

APPENDIX E

WORD LIST

The following listing is Telesensory Speech Systems' basic vocabulary word list for their Series III Speech Module.

Hexadecimal Word Pointer	Word	Hexadecimal Word Pointer	Word
00	zero	12	thousand
01	one	13	plus
02	two	14	minus
03	three	15	time
04	four	16	over
05	five	17	equal
06	six	18	point
07	seven	19	and
08	eight	1A	second
09	nine	1B	degree
0A	ten	1C	dollar
0B	eleven	1D	cent
0C	twelve	1E	pound
0D	thir-	1F	ounces
0E	-teen	20	total
0F	fif-	21	please
10	twenty	22	meter
11	hundred	23	volt

Hexadecimal Word Pointer	Word	Hexadecimal Word Pointer	Word
24	ohm	52	S
25	amp	53	T
26	hertz	54	U
27	stop	55	V
28	high	56	W
29	thank	57	X
2A	is	58	Y
2B	re-	59	Z
2C	low	5A	number
2D	on	5B	left
2E	off	5C	right
2F	in	5D	at
30	out	5E	ready
31	enter	5F	start
32	if	60	set
33	-/z/	61	select
34	-/s/	62	mode
35	low tone	63	next
36	silence 20 ms	64	check
37	silence 40 ms	65	position
38	silence 80 ms	66	test
39	silence 160 ms	67	error
3A	silence 320 ms	68	minute
3B	—	69	date
3C	—	6A	per-
3D	—	6B	call
3E	—	6C	turn
3F	—	6D	try
40	A	6E	code
41	B	6F	rate
42	C	70	fail
43	D	71	continue
44	E	72	wait
45	F	73	switch
46	G	74	hour
47	H	75	move
48	I	76	-crease
49	J	77	put
4A	K	78	million
4B	L	79	liter
4C	M	7A	gram
4D	N	7B	limit
4E	O	7C	—
4F	P	7D	—
50	Q	7E	—
51	R	7F	—

APPENDIX F

INDEX OF MANUFACTURERS

The following is an index of the current manufacturers of speech synthesis components and systems. This list is not complete and is only given so that you, the reader, will have a starting point. You will discover other sources as you continue to work in speech synthesis.

American Microsystems, Inc.
3800 Homestead Road
Santa Clara, CA 95051

General Instrument Corp.
600 West John Street
Hicksville, NY 11802

National Semiconductor Corp.
2900 Semiconductor Drive
Santa Clara, CA 95051

Telesensory Speech Systems
P.O. Box 10099
3408 Hillside Avenue
Palo Alto, CA 94304

Texas Instruments Incorporated
Dallas Technology Center
1001 East Campbell Road
Richardson, TX 75266

Votrax
Div. of Federal Screw Works
500 Stephenson Highway
Troy, MI 48084

INDEX

O

P

R

S

T

U

V

W

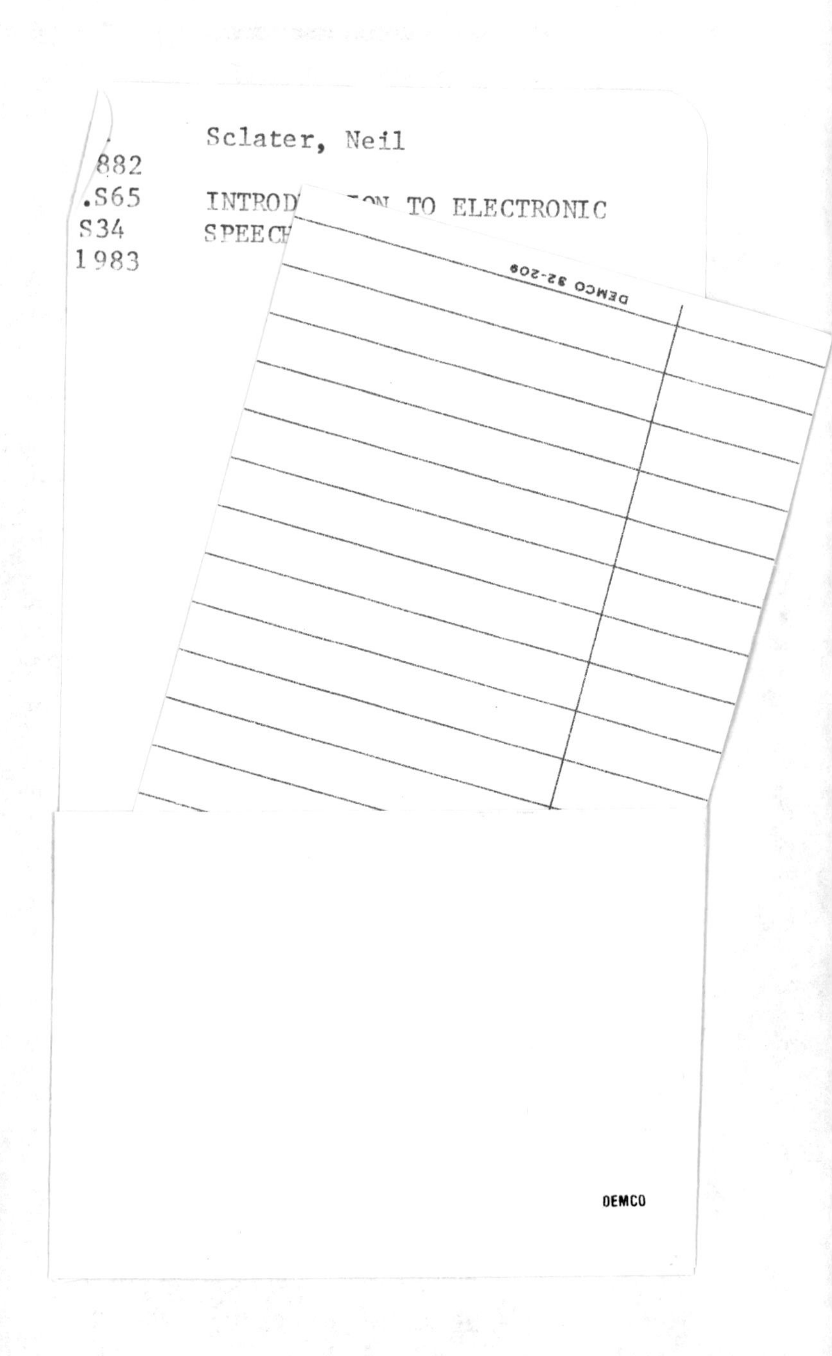
Sclater, Neil
882
.S65
S34
1983
INTROD ON TO ELECTRONIC
SPEECH
DEMCO

READER SERVICE CARD

To better serve you, the reader, please take a moment to fill out this card, or a copy of it, for us. Not only will you be kept up to date on the Blacksburg Series books, but as an extra bonus, **we will randomly select five cards every month, from all of the cards sent to us during the previous month. The names that are drawn will win, absolutely free, a book from the Blacksburg Continuing Education Series.** Therefore, make sure to indicate your choice in the space provided below. For a complete listing of all the books to choose from, refer to the inside front cover of this book. Please, one card per person. Give everyone a chance.

In order to find out who has won a book in your area, call (703) 953-1861 anytime during the night or weekend. When you do call, an answering machine will let you know the monthly winners. Too good to be true? Just give us a call. Good luck.

If I win, please send me a copy of:

____

I understand that this book will be sent to me absolutely free, if my card is selected.

For our information, how about telling us a little about yourself. We are interested in your occupation, how and where you normally purchase books and the books that you would like to see in the Blacksburg Series. We are also interested in finding authors for the series, so if you have a book idea, write to The Blacksburg Group, Inc., P.O. Box 242, Blacksburg, VA 24060 and ask for an Author Packet. We are also interested in TRS-80, APPLE, OSI and PET BASIC programs.

My occupation is ____________________
I buy books through/from ____________________
Would you buy books through the mail? ____________________
I'd like to see a book about ____________________
Name ____________________
Address ____________________
City ____________________
State ____________________ Zip __________

MAIL TO: BOOKS, BOX 715, BLACKSBURG, VA 24060
!!!!!PLEASE PRINT!!!!!

21896